AF713704

WOLF CREEK

635.92
Riker, T. cop 6
The healthy garden book

DISCARDED

Josephine County Library
Grants Pass, Oregon

1000026490

THE HEALTHY GARDEN BOOK

How to Control Plant Diseases, Insects, and Injuries

TOM RIKER

635.92
Ri
Cop 6

STEIN AND DAY/Publishers/New York

JOSEPHINE COUNTY LIBRARY
GRANTS PASS, OREGON

First published in 1979

Copyright © 1979 by Tom Riker

All rights reserved

Designed by Tom Riker

Printed in the United States of America

Stein and Day/Publishers/Scarborough House,

Briarcliff Manor, N.Y. 10510

LIBRARY OF CONGRESS CATALOGING IN PUBLICATION DATA

Riker, Tom.

The healthy garden book.

Includes index.
1. Garden pests--Control. 2. Plant diseases.
3. House plants--Diseases and pests. 4. Gardening
I. Title
SB603.5.R54 635.9'2 78-8407
ISBN 0-8128-2515-2
ISBN 0-8128-6009-8 pbk.

I would like
to dedicate
this book to
my Mother and
Father, Agnes
and Irwin
Riker.

ACKNOWLEDGEMENTS

First, to the many artists of the past for the fine woodcuts and steel engravings that fill these pages.

The United States Department of Agriculture and the Yearbook staff.

The Environmental Protection Agency for the help on pesticide regulations.

Baker Graphics of Westport, Ct.

Quick-Stats and Mike Kusick of Westport, Ct.

My wife Colleen for all her help in the typing and editorial department.

OTHER BOOKS BY TOM RIKER

'The Gardener's Catalogue'
'Food Gardens'
'The Guide to Buying Plants'
'Sex in the Garden'
'City and Suburban Gardens'

'The Hip Pocket Chord Book'
with Erik Darling

DOVETAIL DESIGN WORKS / Tom Riker

—DESIGN FOR A FLOWER GARDEN.

TABLE OF CONTENTS

INTRODUCTION

UNITED STATES & MEXICAN BOUNDARY

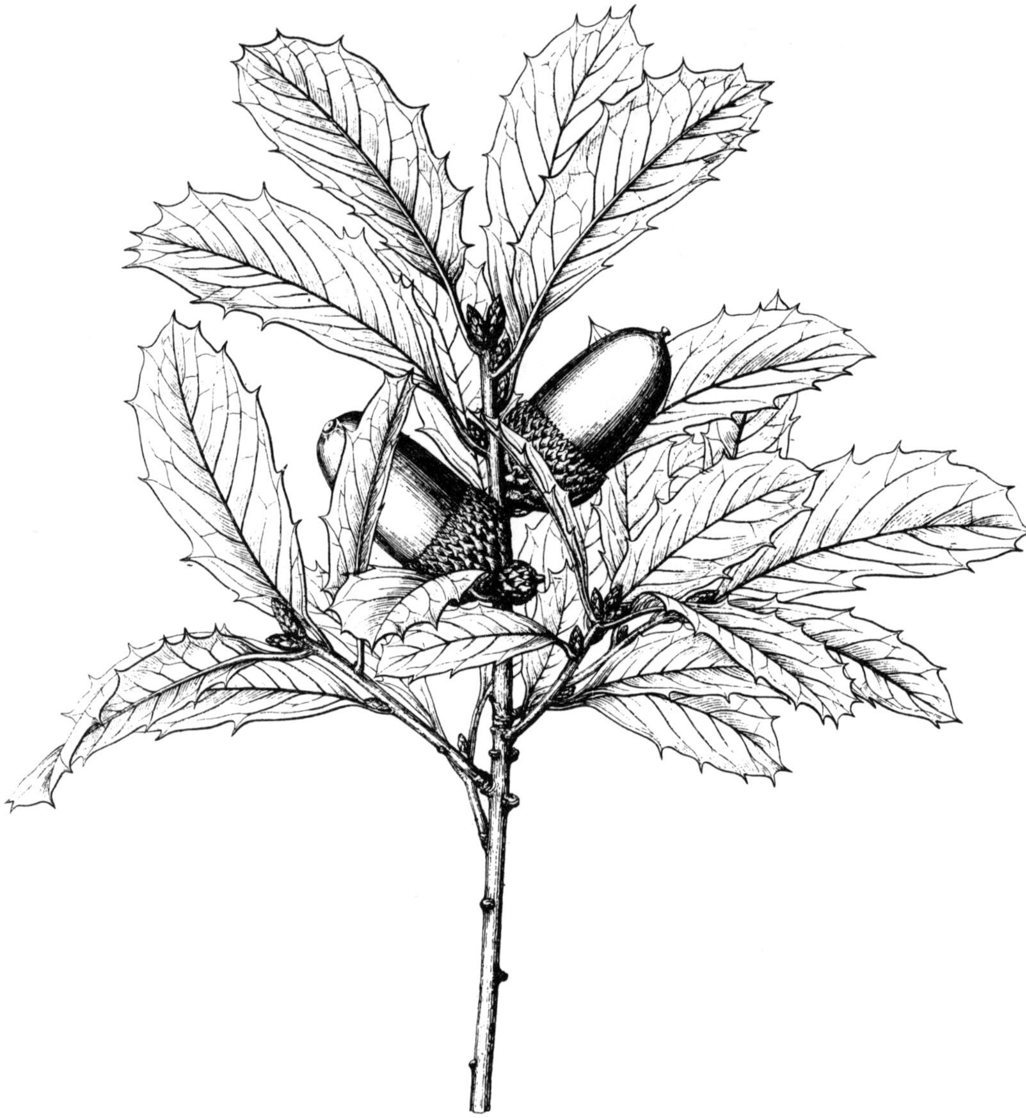

QUERCUS ACUTIDENS.

INTRODUCTION

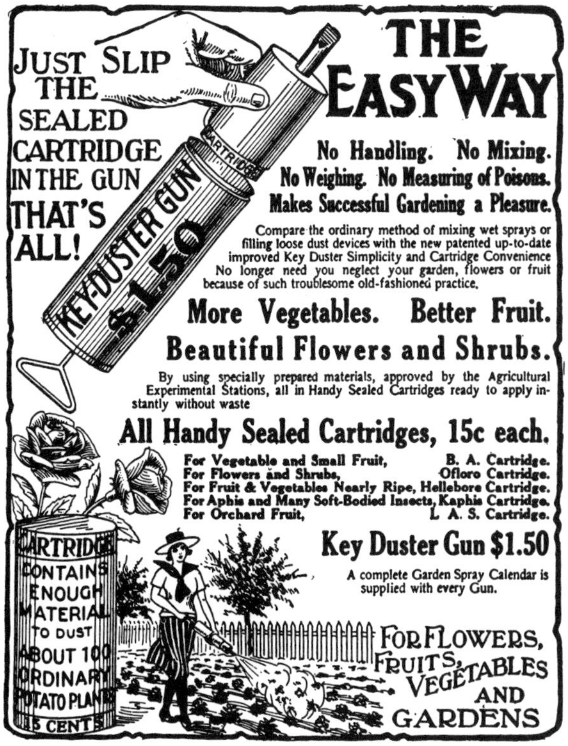

Old seed catalogue advertisement

By no means the least of the difficulties with which the home gardener and plant cultivator has to contend are the many insects which feed on the objects under his care, at times rendering all his other efforts to no avail. To keep a garden completely free from insects and diseases and other problems which professionals generally categorize as "pests", is never an easy task. In some seasons it's an utterly impossible job, but a great deal can be accomplished with a little well-directed care. Prevention is much better than cure and a lot can be done to maintain the garden in good upkeep.

A weedy, uncared for garden corner is a virtual nursery for pests of all kinds. Insects over-winter and diseases rest for the next attack. Rubbish and refuse of last year's crop should never be allowed to collect in any area you plan to cultivate the following year. Any plant part that has been pruned from a diseased plant should be burned or removed from the garden.

Some young plants often suffer more from insect attacks than do mature plants. In such cases the plants should be pushed into vigorous growth as soon as possible. Proper cultivation will help them reach maturity sooner, thus helping them through the most precarious part of their existence.

Birds should be encouraged to frequent your garden. I have always written that birds are our friends in the garden, but few people realize the enormous number of insects they kill. This is especially true during the breeding season, when nearly all young nestlings are fed on insects.

Toads are useful in gardens, too. Many consider these insect eaters ugly, but one toad will do the job of many bottles of chemicals, and that will make those beets and tomatoes you are growing a lot healther for you.

All dead leaves should be collected and removed. I save a few for leaf mold, keeping them in a special bin, but all the rest are bagged. Leaves and old stems of last year's crop of flowers and vegetables are perfect hiding places for insects that over-winter in the soil. After the garden has been cut down for the season, the soil should be turned up to expose any insects. The birds will finish the job.

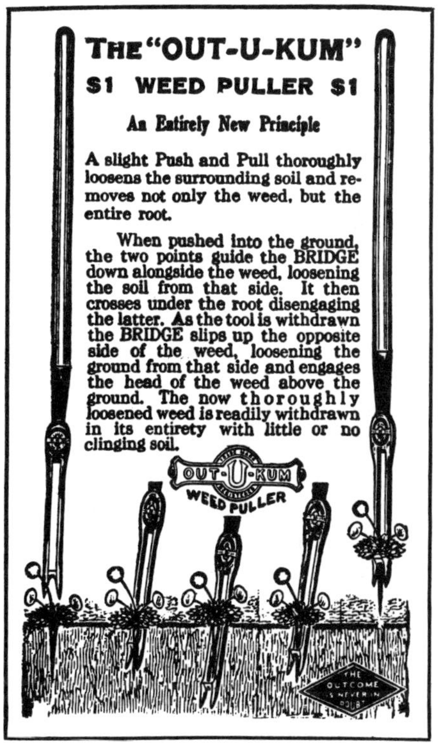

Old seed catalogue advertisement

If you understand the cycles and feeding habits of insects, and the first signs of trouble, you can do without a lot of spraying and use of commercial chemical insecticides. Remedies will be applied in a more intelligent manner if those who use them are acquainted with just a few entomological facts and a few facts concerning plant diseases. A typical adult female insect lays eggs; from these hatch grubs, maggots or caterpillars, depending on the species of insect. These young are both foliage and root feeders and grow quickly. They change their skins several times and finally emerge as insect, moth, butterfly, beetle, bee, wasp, ant or flea that all of us can recognize. Others, such as crickets, grasshoppers, cockroaches, bugs, earwigs, green flies, and scale insects go throught the same metamorphosis, but the changes are less apparent. The young resemble the adult. After the insect reaches full maturity it stops growing. All true insects are similar but because of the shapes and forms they assume, they seem more different than they are. If you look closely you will see three segments: head, thorax or forebody, and body. In a wasp, for instance, they are very clearly defined. The head carries the mouth parts, the feelers or antennae, and the eyes. To the forebody are attached the legs and wings, and the body contains the breathing, digestive, and other internal organs. Every insect has three pair of legs and two pair of wings. Insects breath through pores in their bodies and not through their mouths, or nose as the higher animals do.

Disease problems are more difficult to identify than insects, except when the foliage starts to turn and the plant looks unhealthy. Diseases are caused by bacteria, viruses and fungi. Many times gardeners mistake a chemical spill that affects the plant system as a disease. The fruiting bodies of the fungi are spores and the viruses are microscopic.

Prevention is better than cure, and prevention is often just the application of common sense and an understanding of garden work. If you follow the rules of nature and the few simple steps listed below you will prevent insects from destroying your flowers, vegetables, fruit trees and potted plants.

Cicada.

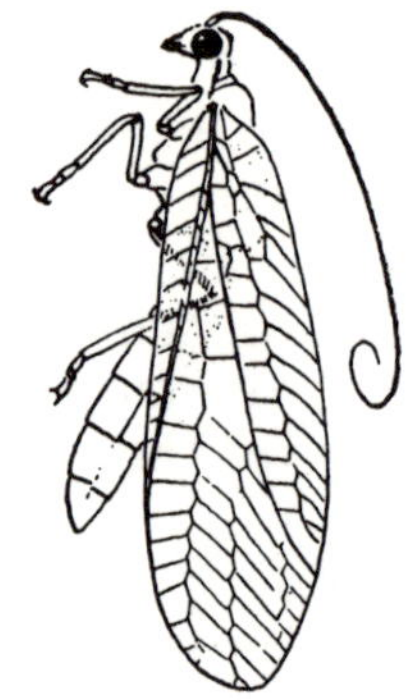

Lacewing.

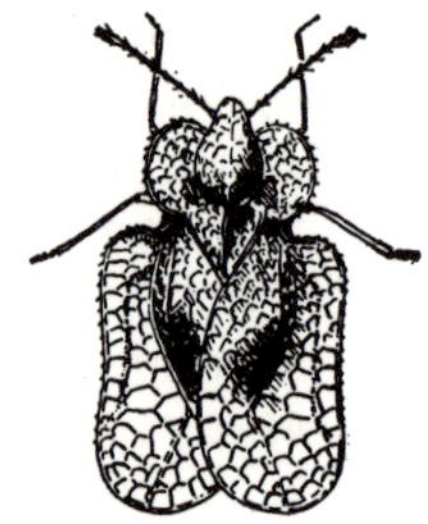

Lace bug.

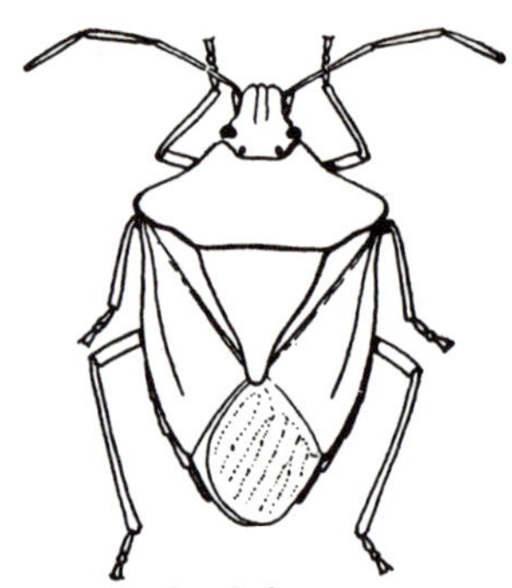

Stink bug.

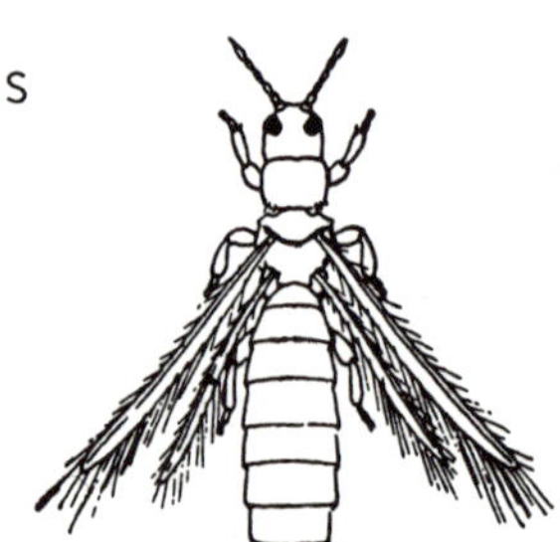

Thrips.

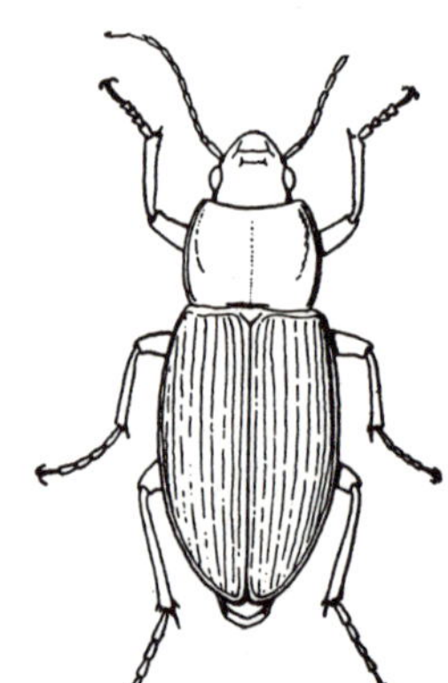

Ground beetle.

Lady beetle.

1. Plant in a well drained fertile soil. This means a soil that is not hardpan. A light, sandy soil with a good supply of organic matter is best. Plants need space, too, to develop root systems.

2. At the beginning of the growing season cultivate into the soil a conditioner such as peatmoss or cow manure. If the soil needs lime, use an agricultural lime. You can buy soil testing kits to meet your requirements at your local garden store. If the site is well prepared before the plants arrive or when you are ready to transplant them from the cold frame to the garden, you are already a step ahead.

3. Vegetables need sun, so plant in a southern exposure. Dampness and wet areas attract fungi.

4. Maintain a regular fertilizing program throughout the season.

5. Do not plant shrubs, trees, or any plants that are not suited to the soil or region. Make sure the plants you buy are hardy for your climate. You would do well to buy from a local nursery or just keep in mind, when buying mail-order plants, where they are coming from and where they have to grow.

RHODODENDRON CATAWBIENSE.

6. Keep out weeds; they attract insects and diseases.

7. Plant to try and attract birds; they're natural insecticides.

8. Employ companion planting. Use marigolds and onions in the rose beds, for example, to keep plants free of aphids. More on companion planting and natural controls later in the book.

9. Buy disease resistant varieties. Buy from established, reputable dealers.

10. Plant seeds in trays with sterilized soil. This will prevent damping-off (see Chapter 4).

SWEET-SCENTED SHRUB.
(CALYCANTHUS FLORIDUS.)

11. Keep the garden clean of all plants of the past growing season. Insects and diseases over-winter in leaves, on stems, and on other plant parts.

12. Understand early symptoms of diseases and insect damage. The following two chapters will tell you what to look for.

13. Keep a note book of garden dates and keep a garden log table. Know where everything is planted.

14. If you are a beginner, ask questions of people who have gardened in the area before. Each region has its own problems.

15. Read some books on gardening and get to know your local extension service agent.

To repeat, natural insect and disease controls are preferable to chemical controls, but for those who want to use the latter, I have detailed the proper procedure. A lot of trouble is caused each year by over-spraying, over-fertilizing, and general misuse of chemicals. There is a bewildering array of chemicals available. If you are not sure what your problem is, or if you believe the information on the package is not correct, ask a professional for help. You probably buy your garden supplies from a local garden center. Do not be afraid to ask questions of the dealer. If you buy in a hardware store, where the salesman is an expert in tools and nails, you will probably not find the help you need. But if you buy garden supplies in a garden center, expert information should also be available. If it is not, go down the street or road to another dealer.

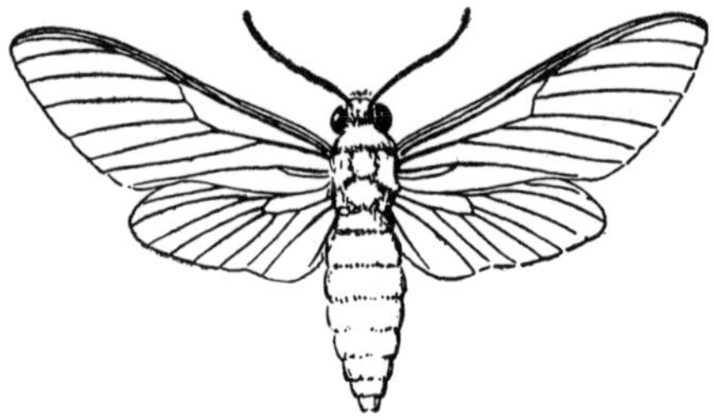
Moth.

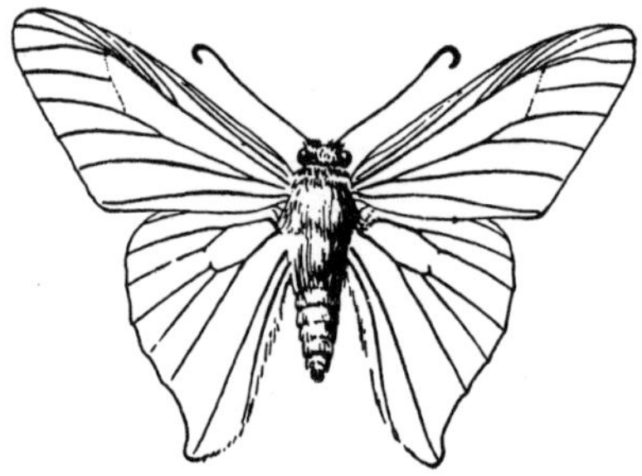
Skipper.

Butterfly.

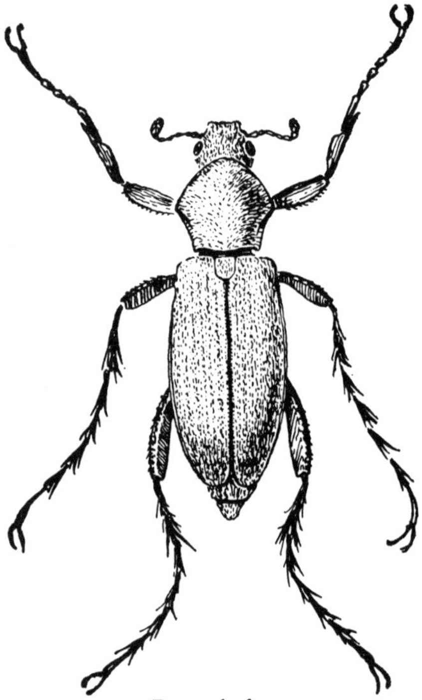
Rose chafer.

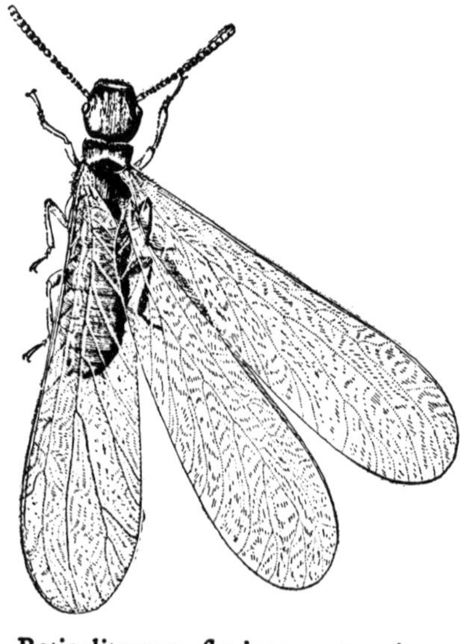
Reticulitermes flavipes, a termite.

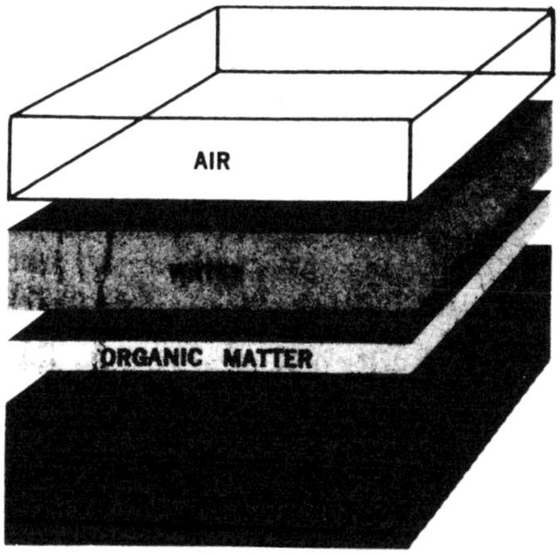

RELATIVE COMPOSITION OF A SOIL WITH GOOD STRUCTURE

Common Sources of Fertilizer Nitrogen for Gardens

Name	*Grade*	*Nitrogen*
Chemical		*Percent*
Ammonium nitrate	33–0–0	33.5
Ammonium sulfate	21–0–0	21
Sodium nitrate	16–0–0	16
Urea	45–0–0	45
Urea formaldehyde	46–0–0	46
Organic		
Dried blood	11–0–0	13.3
Castor pomace	5–2–1	5.5
Cottonseed meal	6–3–0	7.0
Dried manure	varies	varies
Activated sewage sludge	6–3–0	varies

Common Sources of Potassium Fertilizers

Material	*Grade*	*Potash*
		Percent
Muriate of potash	0–0–60	60 – 63
Sulfate of potash	0–0–50	50 – 53
Sulfate of potash-magnesia	0–0–22	22 – 23
Nitrate of potash	13–0–44	44 – 46

Common Sources of Phosphate for Gardens

Material	*Grade*	*Available phosphoric acid*
		Percent
Rock phosphate	—	
Normal superphosphate	0–20–0	20
Triple superphosphate	0–45–0	45
Ammonium phosphate	variable [1]	14 – 53
Steamed bonemeal	2–27–0 [2]	18 – 34

[1] 11–48–0, 13–29–0, 16–20–0, 21–53–0, and 27–14–0.

[2] Averaged values from State fertilizer control officials.

Primary Nutrients Contained in Inorganic Fertilizers

Fertilizer	Formula	Percent Nitrogen N	Available: Percent Phosphorus P_2O_5	Available: Percent Potassium K_2O
Ammonium nitrate	NH_4NO_3	33.5	0	0
Ammonium sulfate	$(NH_4)_2SO_4$	21	0	0
Calcium nitrate	$Ca(NO_3)_2$	16	0	0
Nitrogen solutions	(Varies)	20–50	–	–
Sodium nitrate	$NaNO_3$	16	0	0
Urea	$CO(NH_2)_2$	45	0	0
Ammonium phosphate	$NH_4H_2PO_4$ mostly	11	48	0
Diammonium phosphate	$(NH_4)_2HPO_4$	18	46	0
Basic slag	Ca, Mg, Al silicates high in phosphates	0	8	0
Rock phosphate	$3Ca_3(PO_4)_2 \cdot CaF_2$ mostly	0	5	0
20 percent superphosphate	$CaH_4(PO_4)_2$ and $Ca_2H_2(PO_4)_2$	0	20	0
Concentrated superphosphate	$CaH_4(PO_4)_2$	0	45	0
Superphosphoric acid, polyphosphate	H_3PO_4 and $H_4P_2O_7$	0	76	0
Green sand (Glauconite)	$KFeSi_2O_6 \cdot nH_2O$	0	1	6
Muriate of potash	KCL	0	0	60
Potassium sulfate	K_2SO_4	0	0	50
Potassium magnesium sulfate	$K_2SO_4 \cdot 2MgSO_4$	0	0	21
Potassium nitrate	KNO_3	13	0	44

—HARDY HYDRANGEA (*Hydrangea Paniculata, grandiflora*).

GARDEN PESTS AND WHAT THEY DO

GARDEN PESTS

Five agents harmful to plants commonly characterized as "pests".

Insects and arachnids

Snails and slugs

Pest animals (moles, mice and rats)

Weeds

Plant diseases

You probably will find some pests that you cannot idenify. Never try to control a pest until you are sure what it is.

SOFT SCALE ON ORANGE SPRAY.
References: 1, female, natural size; 2, female, magnified.

Insects and Arachnids

Many kinds of insects are pests. They may:

. Feed on leaves

. Tunnel or bore in stems, stalks and branches.

. Feed on and tunnel in roots.

. Suck the sap from leaves, stems, roots, **fruits and flowers.**

. Carry plant and animal diseases.

. Feed on or in man and other animals.

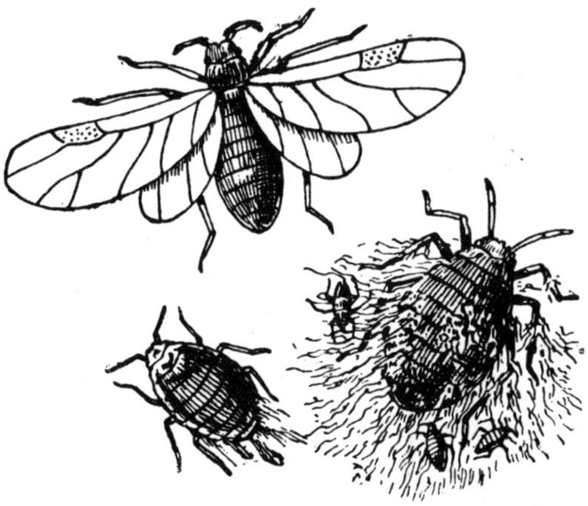

WOOLLY APHIS OR AMERICAN BLIGHT.
References: *Upper figure*, winged female; *lower left-hand figure*, wingless female; *lower right-hand figure*, viviparous female and young in wool.

All adult insects have two things in common; they have six jointed legs and three body segments. Almost all insects change shape, form, and size during their lives.

You can tell one insect from another by looking at the wings and mouthparts. Some insects have two wings, others have four. The wings vary in shape, size, thickness, and structure.

Insects feed in different ways. Those with chewing mouthparts have toothed jaws that bite and tear food. Insects with piercing-sucking mouthparts have a long beak which they force into a plant or animal to suck out the fluids or blood.

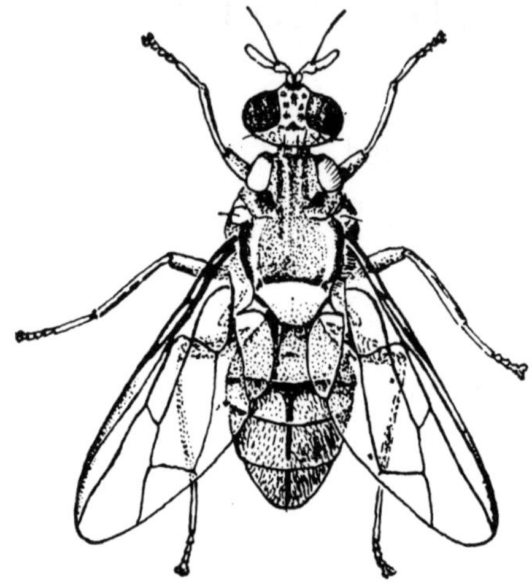

Oriental fruit fly.

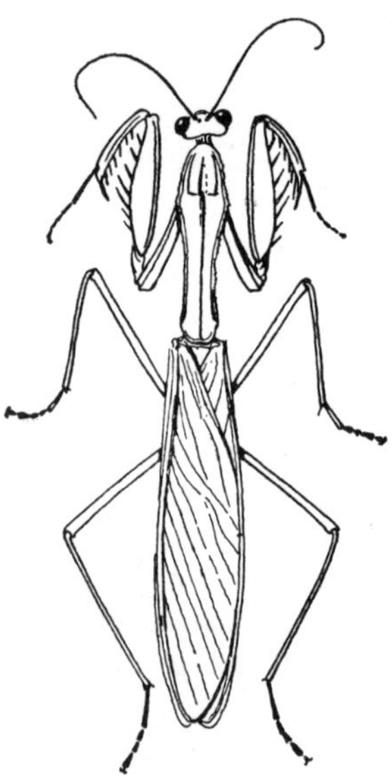

Praying mantid.

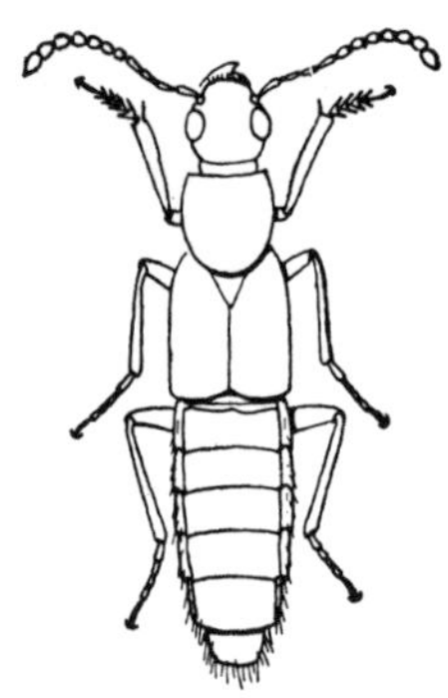

Rove beetle.

Weevil.

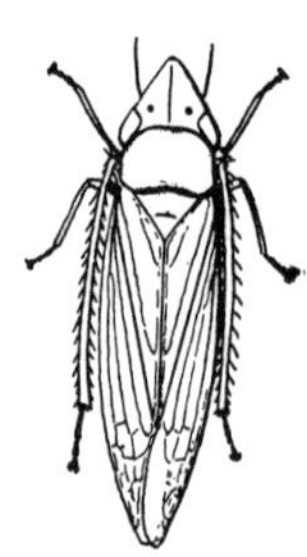

Leafhopper.

Sawfly.

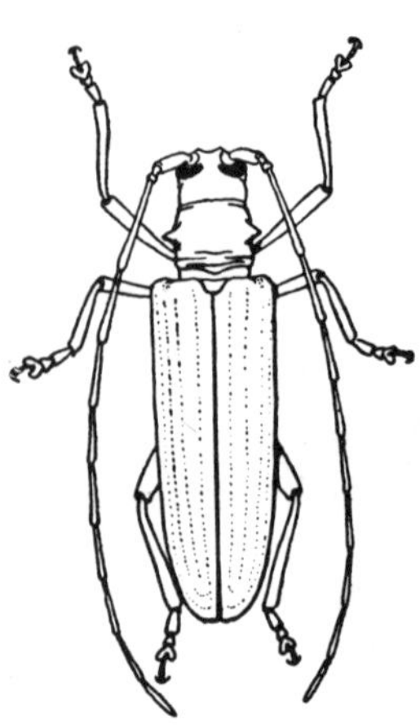

Longhorned beetle.

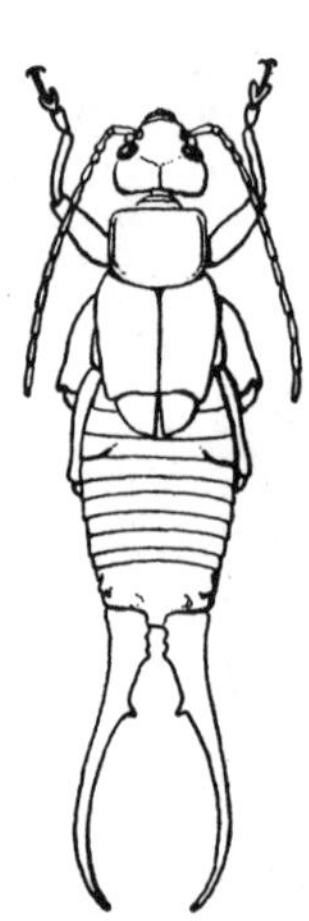

Earwig.

Mediterranean fruit fly.

Mites, Ticks, and Spiders 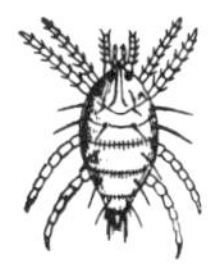

Mites, ticks, and spiders are arachnids--closely related to insects. The main differences are that the adults have eight jointed legs instead of six and have two body segments. They do not have wings.

Snails and Slugs

Snails and slugs are members of a large group of animals called mollusks. Snails have a hard shell; slugs have no shell. They feed on plant foliage, and are pests in lawns, landscape plantings, and greenhouses.

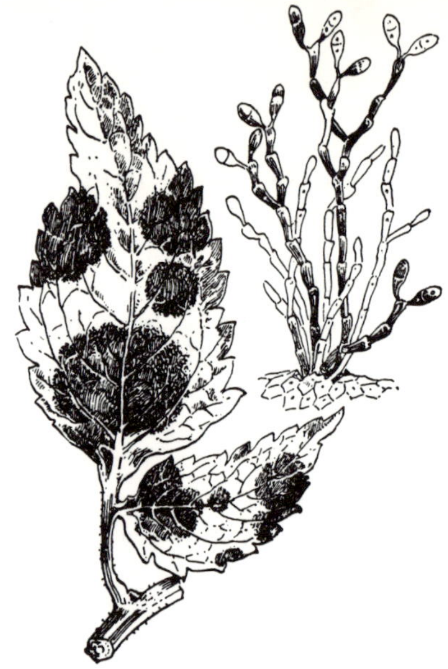

Tomato Leaf-Rust (*Cladosporium fulvum*).

Plant Diseases

Professional gardeners broadly characterize a plant disease as any harmful condition that makes a plant different from a normal plant. There are two main kinds of plant diseases.

Diseases caused by non-living agents:

The causes include such things as frost, air pollution, and drought. These diseases cannot be passed from one plant to another.

Diseases caused by living things:

The most common causes of disease agents and examples of the diseases they cause are:

. Fungi (scab and rot)

. Bacteria (blights, wilts, and scabs)

. Viruses (mosaics)

. Nematodes (root rots and cysts)

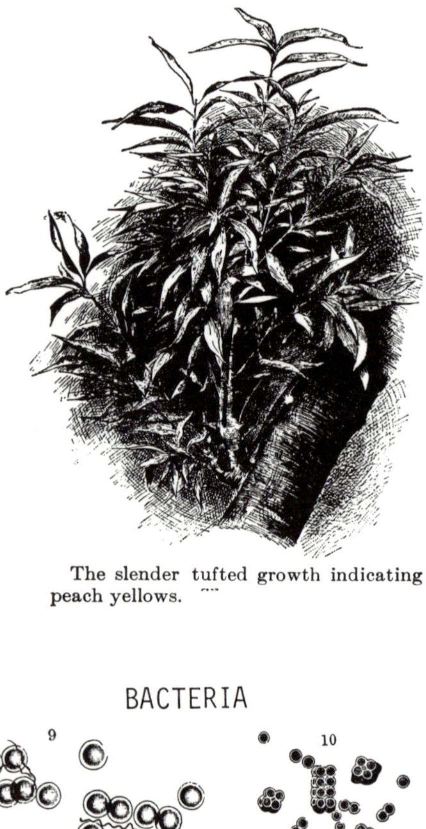

The slender tufted growth indicating peach yellows.

BACTERIA

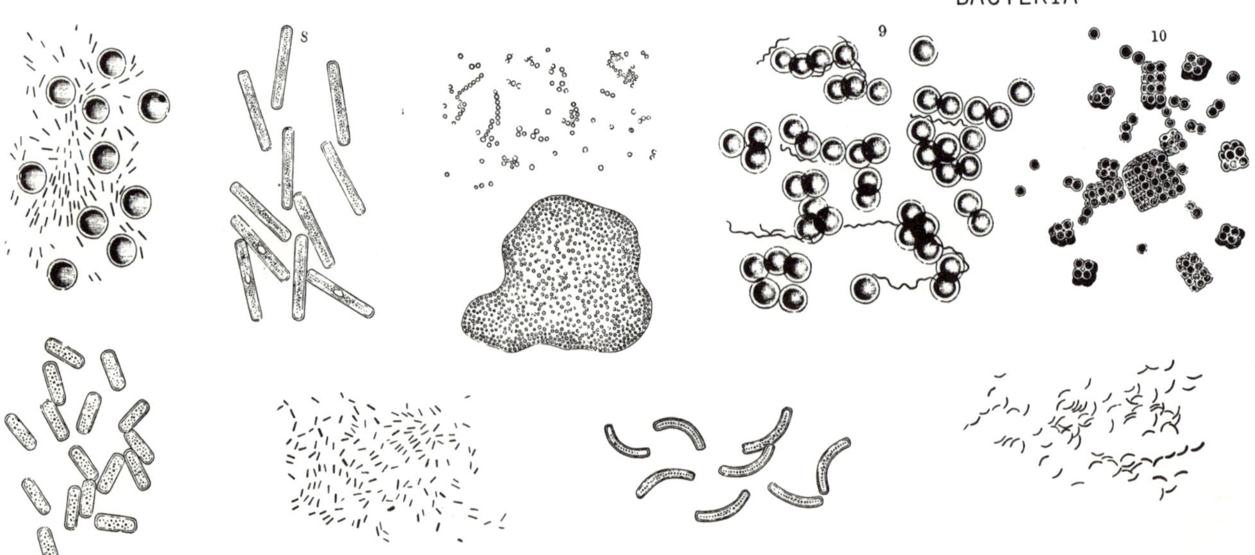

If you have idenified a pest, and you know how it grows, spreads, and does its damage, you can plan how to control it. Using pesticides is only one of many ways to control pests; a combination of methods is basic to the process. Be sure to consider what other methods might work before you decide to apply a pesticide. You can minimize pest problems by:

. Encouraging pests' natural enemies

. Planting crop varieties that resist pests

. Destroying crop residues

. Practicing good manure management

. Clean plowing and cultivation

Pest control is necessary only when the pest is causing more damge than it is reasonable to accept. Even though a pest is present, it may not do much harm. It may cost more to control the pest than you may lose because of the damge it causes.

Weeds

A weed is simply a plant out of place. Weed plants take food out of the ground that normally would be used by the plants under cultivation. Some perennial weeds are also host plants of diseases.

Pest Animals

Any of the digging and tunneling rodents are a problem in the garden. You can trap them or use poison bait. I personally like many of the field mice and moles, but they are a problem and you can eliminate them if you wish.

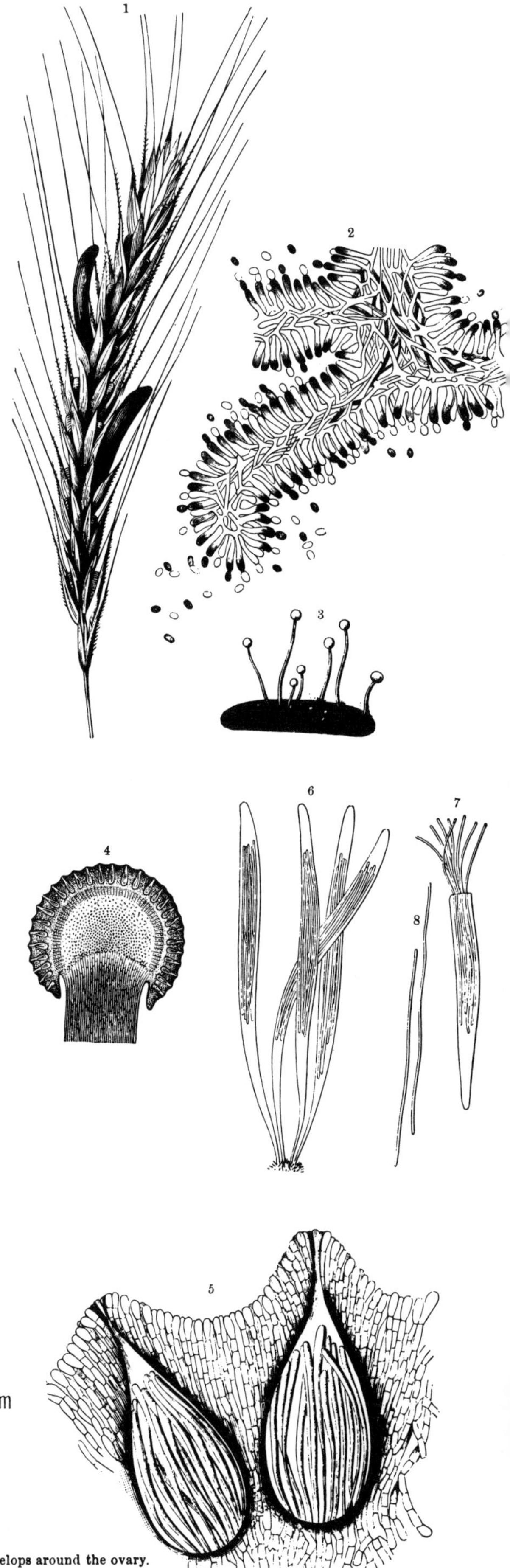

–The Ergot of Rye, *Claviceps purpurea.*

1 Ear of Rye showing two sclerotia of the Fungus. 2 Conidia arising from the mycelium which develops around the ovary. 3 Stalked stromata arising from the sclerotium. 4 Longitudinal section through the head of a stroma showing the perithecia at the edge. 5 Vertical section through two perithecia showing the asci. 6 Asci. 7 Ascus liberating its filamentous spores. 8 Filamentous spores. 1, 3 nat. size; 2 ×200; 4 ×40; 5 ×50; 6 and 7 ×700; 8 ×750. (Partly after Tulasne.)

PESTICIDES

This compressed-air sprayer holds 2 gallons of spray, and it operates at about 50 pounds of pressure, which is obtained by a plunger-type handpump. The handle of the pump also serves as the handle of the sprayer.

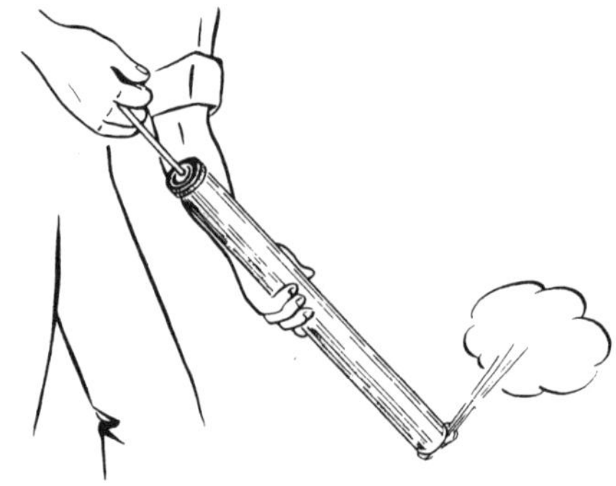

This plunger-type duster is ideal for treating individual plants and small areas.

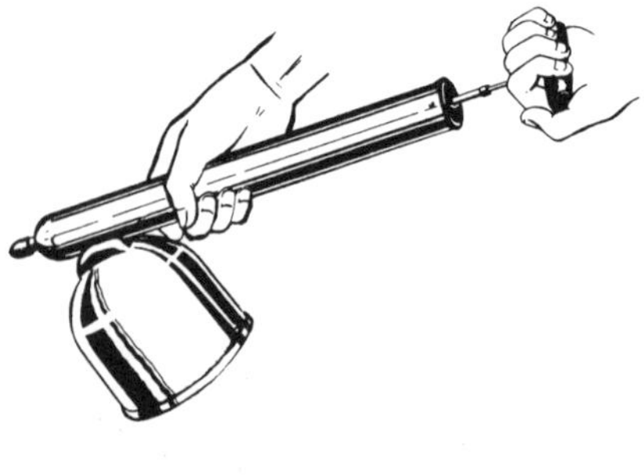

This hand atomizer has a copper spray chamber. Some hand atomizers come with removable glass chambers that resemble fruit jars.

How Pesticides Work

Pesticides can be grouped according to what they do. Many work in more than one way. Read the label to find out what the one you are using will do.

Some of the ways pesticides work are:

Contacts: kill pests simply by contacting them.

Stomach poisons: kill when swallowed.

Systemics: kill pests by being taken into the blood of the animal or sap of the plant upon which the pest is feeding.

Translocated herbicides: kill plants by being absorbed by leaves, stems, or roots and moving throughout the plant.

Fumigants: gases which kill

INSECTICIDES AND FUNGICIDES

You'll find it useful to understand the differences among the various chemical compounds that make up all the formulations that are used as pest controls. Before we look at the listing of chemicals and the proper way to apply them, you should consider the precautions. At this writing more and more compounds are coming off the commerical market shelves and, sorry to say, more are going on. The latest information comes from the Environmental Protection Agency, now in complete control of pesticide regulations. At one time the Department of Agriculture controlled chemicals, but the amount of information on this subject is so staggering that an agency concerned solely with the protection of our welfare has taken over the vast job of sorting out the good from the bad. A chemist one time told me if you can smell it, it probably will do some harm. If you've ever looked at a trade list of available agriculture chemicals, you'll understand the problems facing this agency.

The hazards and risks of using chemicals to eliminate pests in the home garden have increased in recent years due to widespread usage. If you have any questions concerning the regulations of pesticides, you can write or call your representative in Congress and ask for a copy of the Federal Insecticide, Fungicide, and Rodenticide Act (public law 92-516), and a book published by the EPA in 1976 amending the above law. Investigation continues to uncover new dangers of pesticides; a listing of the cancelled pesticides is now available from the EPA.

Most home gardeners still use chemicals to control pests and a detailed outline of the misuse, danger and proper use is appropriate to begin any discussion of those controls. As long as the general public has been sold on poisons to control garden pests and diseases, everyone should understand how they work.

Pesticides used improperly can injure your health and your pets, as well as your plants. Most of the troubles that occur around the home, reports from poison control centers indicate, could have been avoided if the pesticide had been applied properly-- if the user had read the label before spraying or dusting the garden.

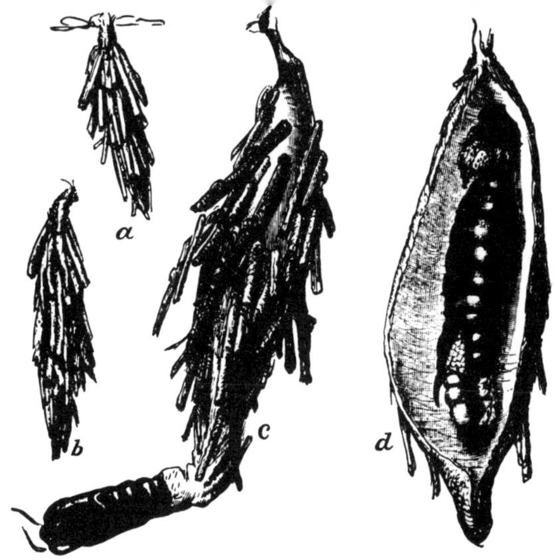

—Bagworm at (*a*, *b*, *c*) successive stages of growth. *c*, male bag; *d*, female bag—natural size (original).

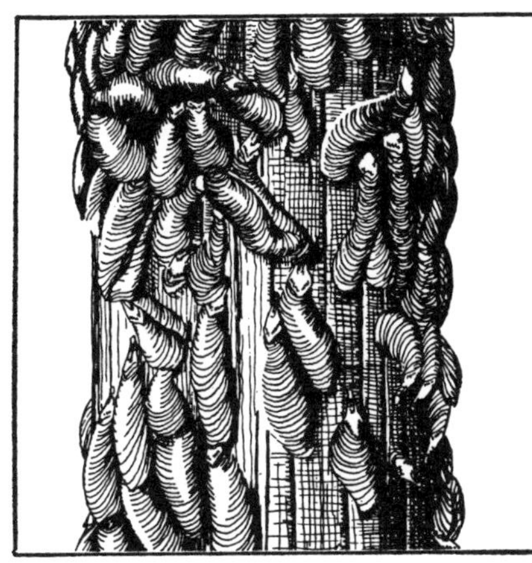

Oystershell scale.

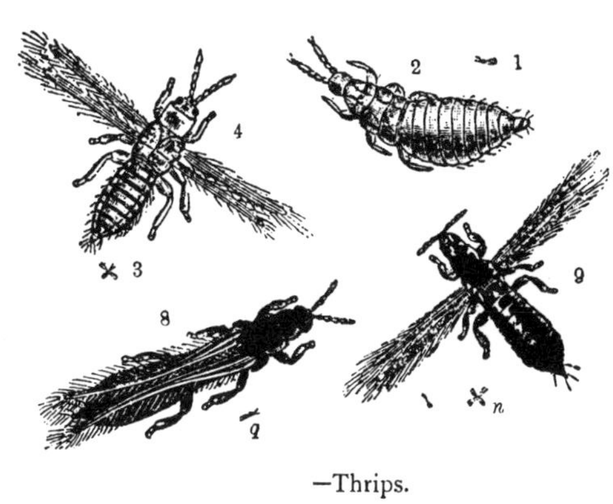

—Thrips.
1 and 2. Immature Thrips. 3 and 4. Mature Thrips. 8 and 9. Other species of Thrips.

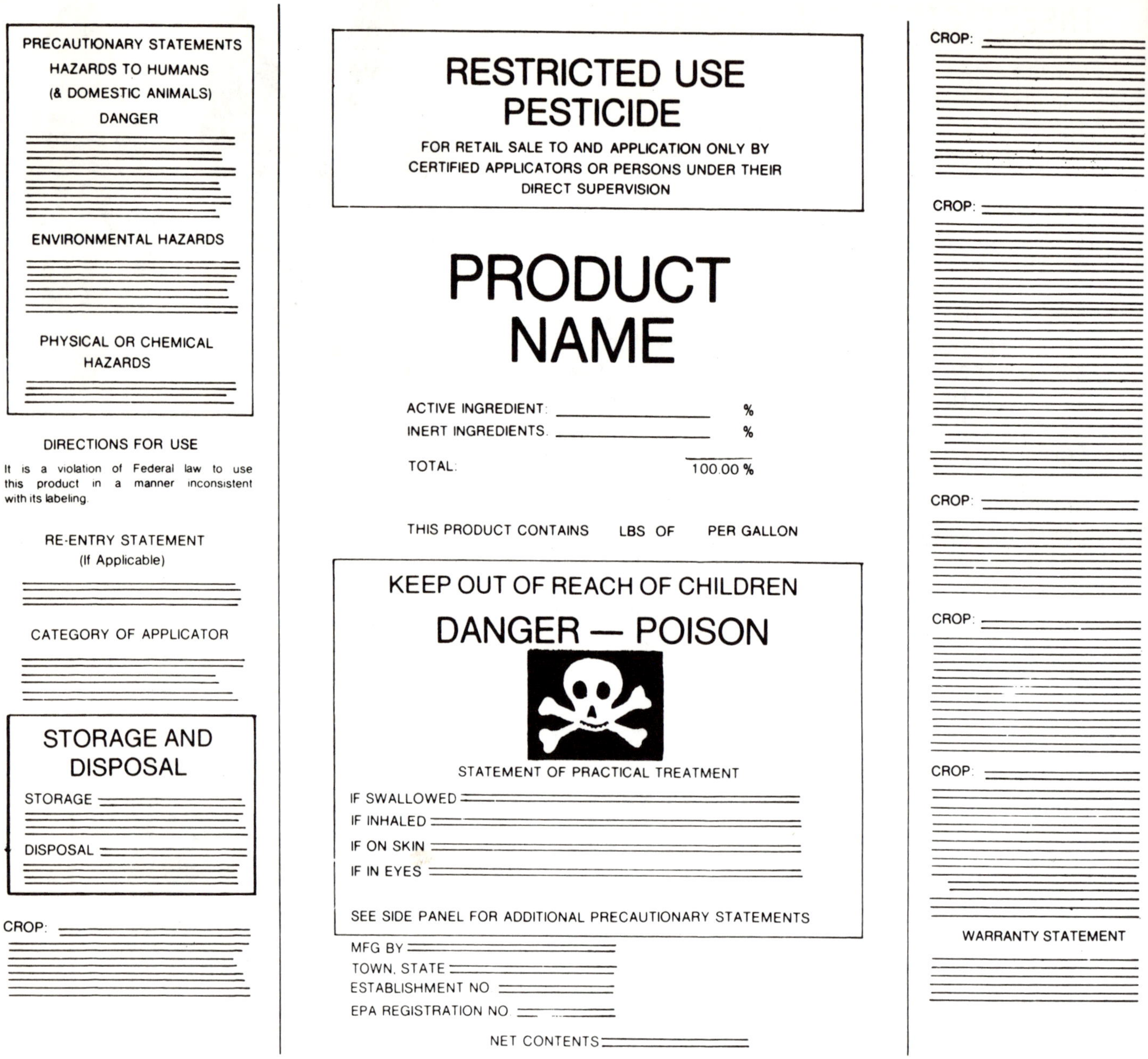

PRECAUTIONARY STATEMENTS
HAZARDS TO HUMANS
(& DOMESTIC ANIMALS)
DANGER

ENVIRONMENTAL HAZARDS

PHYSICAL OR CHEMICAL HAZARDS

DIRECTIONS FOR USE
It is a violation of Federal law to use this product in a manner inconsistent with its labeling.

RE-ENTRY STATEMENT
(If Applicable)

CATEGORY OF APPLICATOR

STORAGE AND DISPOSAL
STORAGE
DISPOSAL

CROP:

RESTRICTED USE PESTICIDE
FOR RETAIL SALE TO AND APPLICATION ONLY BY CERTIFIED APPLICATORS OR PERSONS UNDER THEIR DIRECT SUPERVISION

PRODUCT NAME

ACTIVE INGREDIENT: %
INERT INGREDIENTS: %
TOTAL: 100.00 %

THIS PRODUCT CONTAINS LBS OF PER GALLON

KEEP OUT OF REACH OF CHILDREN
DANGER — POISON

STATEMENT OF PRACTICAL TREATMENT
IF SWALLOWED
IF INHALED
IF ON SKIN
IF IN EYES

SEE SIDE PANEL FOR ADDITIONAL PRECAUTIONARY STATEMENTS

MFG BY
TOWN, STATE
ESTABLISHMENT NO
EPA REGISTRATION NO

NET CONTENTS

CROP:
CROP:
CROP:
CROP:
CROP:

WARRANTY STATEMENT

Lets start with that label. After the nurseryman or garden center operator sells you a bottle or box of chemicals, read the label to learn the correct mixing and dosage. New labels, as mandated by law, must contain the following:

1. Brand name

2. Type of formulation

3. Ingredient statement

4. Common name (many compounds have very long chemical names but the manufacturers and distributors often use common chemical names to make identification easier for the layman. These are not trade names.

5. Net contents (quarts, pounds, or other units of measure.)

6. Name and address of manufacturer.

7. Establishment number (the factory that made the chemical). It does not have to be on the label but must be on the bottle or container.

8. Registration number (federal number, by law)

9. Precautionary statements:
   a. Hazards to human and domestic animals.
   b. Environmental hazards (for example: "toxic to bees," "do not apply where run-off is possible")

10. Physical and chemical hazards
    a. fire hazards
    b. explosion or chemical hazards

11. Signal words and symbols: Many pesticides are hazardous to people; you can tell how toxic they are by the signs and symbols on the label. (For example; Danger; Highly toxic, a taste to a tablespoon is enough to kill an average person- or: Warning Moderately toxic and Caution: Low Toxicity.)

12. All products must carry the instruction "Keep out of the reach of children."

13. Statement of practical treatment: This is a brief medical statement telling the user if the compound is dangerous to your skin, your eyes, or is dangerous to inhale. All labels should include first aid measures. You should always take the label to the doctor if trouble does occur.

14. Directions for use:

    a. Pests the product controls.

    b. The plants on which it may be used.

    c. Whether the product is for commercial or general use.

    d. How to apply the product.

    e. How much to use.

    f. Where the product should be applied.

    g. When the product should be applied.

15. Misuse statement: (It can be a violation of federal law if the product is used in any manner except that stated on the label.)

16. Reentry statement: (Tells you how soon it is safe to enter the area just sprayed or dusted.)

17. Category of Applicator: (If required by the product, certain products can only be used under special conditions by professional applicators.)

Aerosols (A)

These are low concentrate solutions, usually applied as a fine spray or mist indoors. Some are sold in pressurized cans.

Liquified Gases

These fumigant formulations turn into a gas when they are applied. Some of them have to be packaged in pressure containers.

Liquid Formulations

Emulsifiable Concentrates (EC or E)

An emulsifiable concentrate can be mixed with water to form an emulsion in your spray tank.

Ultra Low Volume Solutions (ULV)

These formulations may contain only the active ingredient itself. They require special application equipment.

Solutions (S)

These formulations are ready to use. They are often used on livestock and in barns.

Flowables (F or L)

A flowable can be mixed with water to form a suspension in your spray tank.

Dry Formulations

Dusts (D)

Dust formulations are made by adding the active ingredient to a fine inert powder. Dusts must be used dry.

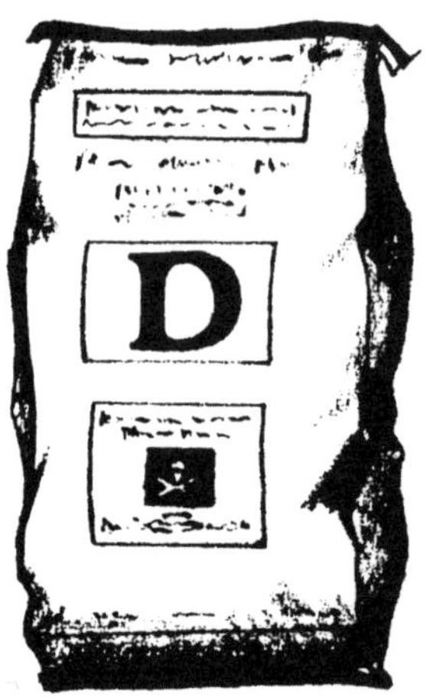

Baits (B)

A bait formulation is made by adding the active ingredient to an edible or attractive substance.

Soluble Powders (SP)

A soluble powder formulation is made from an active ingredient that dissolves in water.

Granules (G)

Granular formulations are made by adding the active ingredient to coarse particles (granules) of some inert material. Granule particles are much larger than dust particles.

Wettable Powders (WP or W)

Wettable powder formulations are made by combining the active ingredient with a fine powder and a wetting agent. They look like dusts, but they are made to mix with water. These formulations need continuous agitation to maintain a suspension.

Here are the types and uses of pesticides:

Insecticide: controls insects and other related pests such as ticks and spiders.
Miticide: controls mites.
Acaricide: controls mites, ticks and spiders.
Nematicide: controls nematodes.
Fungicide: controls fungi.
Bactericide: controls bacteria.
Herbicide: controls weeds.
Rodenticide: controls rodents.
Avicide: controls birds.
Piscicide: controls fish.
Molluscicide: controls mollusks, such as slugs and snails.
Predacide: controls pest animals.
Repellent: keeps pests away.
Attractant: lures pests.
Plant Growth Regulator: stops, speeds up, or otherwise changes normal plant processes.
Desiccants and Defoliants: harvest-aid chemicals used to remove or kill leaves and stems.
Antitranspirant: Reduces water loss from plants.

Weights and Measures

Weights
16 ounces = 1 pound
1 gallon water = 8.34 pounds

Liquid Measure
1 fluid ounce = 2 tablespoons
16 fluid ounces = 1 pint
2 pints = 1 quart
8 pints = 4 quarts = 1 gallon

Length
3 feet = 1 yard
16 1/2 feet = 1 rod
5,280 feet = 320 rods = 1 mile

Area
9 square feet = 1 square yard
43,560 square feet = 160 square rods = 1 acre

Speed
1.466 feet per second = 88 feet per minute = 1 mph

Volume
27 cubic feet = 1 cubic yard

Pesticide formulation	Amount of insecticide needed per gallon of spray — Desired concentration 0.03%	0.5%	1.0%
carbaryl (Sevin) 50% W.P.	..	4 tbsp.	8 tbsp.
chlordane 45% E.C.	..	8 tsp.	5 tbsp.
chlordane 72% E.C.	..	4 tsp.	8 tsp.
chloropropylate (Acaralate) 2E.	1 tsp.	..	..
diazinon (Spectracide) 25% E.C.	..	5 tbsp.	10 tbsp.
dicofol (Kelthane) 18.5% W.P.	2 tsp.	..	..
dicofol (Kelthane) 18.5% E.C.	1½ tsp.	..	..
Dursban 2E.	..	..	3 tbsp.
malathion 50–57% E.C.	..	7 tsp.	4½ tbsp.

(tbsp. = tablespoon; tsp. = teaspoon)

NAMES OF INSECTICIDES

Below is a list of the common names of insecticides used in the preceding tables, followed by the commercial trade name and the chemical name. Some products may be available under a variety of trade names not listed below. Be sure to read the label. The label on the container always lists these products by the common name or chemical name.

Common Name	Trade Name	Chemical Name
carbaryl	Sevin	1-naphthyl methylcarbamate
chlorpyrifos	Dursban	O, O-diethyl O-(3,5,6-trichloro-2-pyridl) phosphorothioate
deet	Off, Kik	N, N-diethyl-m-toluamide
diazinon	Spectracide	O, O-diethyl O-(2-isopropyl-4-methyl-6-pyrimidyl) phosphorothioate
dichlorvos	Vapona, DDVP	2,2-dichlorovinyl dimethyl phosphate
dicofol	Kelthane	4,4′-dichloro-a-(tri = chloromethyl) benzhydrol
dimethoate	DeFend, Cygon	O, O-Dimethyl S-(N-Methyl carbamoyl methyl) phosphorodithioate
ethyl hexanediol	6-12, Rutgers 612	2-ethyl-1, 3-hexanediol
malathion	Cythion	diethyl mercaptosuccinate, S-ester with O, O-dimethyl phosphorothioate
propoxur	Baygon	O-isopropoxyphenyl methylcarbamate
pyrethrin		principally from plant species *Chrysanthemum cinariaefolium*

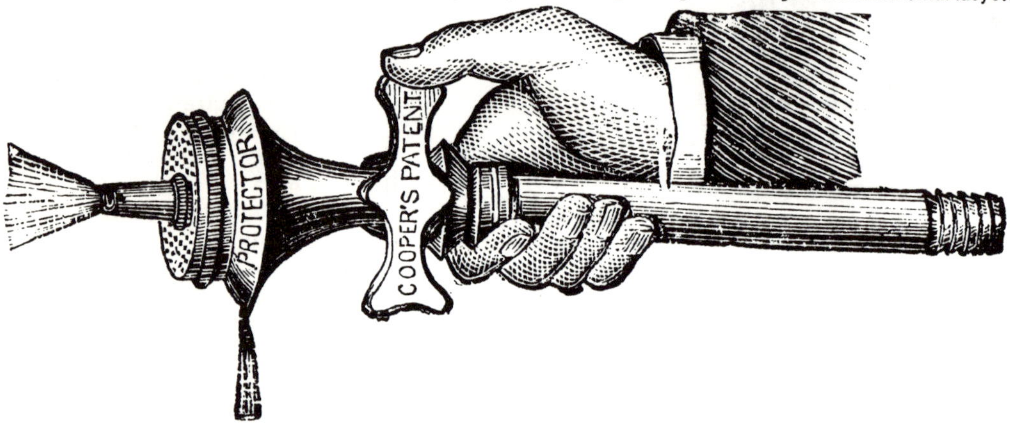

SAFE USE OF PESTICIDES

The two basic reasons to use pesticides safely are to keep yourself and other folks from being poisoned, and to avoid harming the plants and animals in the garden. The symptoms of pesticide poisoning vary with the compound and the contact.

1. A mild poisoning will usually produce:
   Fatigue
   Headache
   Dizziness
   Blurred vision
   Inordinate sweating
   Nausea and vomiting
   Stomach ache or diarrhea

2. A moderate poisoning will usually produce:
   Inability to walk
   Weakness
   Chest discomfort
   Muscle twitches
   Constriction of the pupil of the eye
   Earlier symptoms become more severe

3. Severe poisoning will usually produce:
   Unconsciousness
   Difficulty in breathing
   Same symptoms as moderate but more severe
   Death if not treated

If this list troubles you, it should. This information is from the United States government. Use of chemicals may not be worth the possible hazard. I can only recommend that you use chemicals as sparingly as possibe.

If you still want to use chemicals use protective clothing: long sleeved shirts, gloves, hat and rubber boots. If you are spraying in the air, at an apple tree for example, use a respiratory device over your mouth and nose. Protect yourself.

HOW PESTICIDES CAN HARM YOUR GARDEN

Pesticide spray mists can drift to nearby gardens and other areas that will not tolerate the agent. You may kill bees and other pollinators if you spray an insecticide in an area in which pollination is taking place. You may kill wildlife in streams and ponds if you spill chemicals or if you throw away a container near a waterway. If you spray the wrong pesticide on the wrong plant you may kill it. Herbicides like 24D or Silvex are only for use on lawns, or clearing away brush. These strong chemicals should not be used to kill weeds around cultivated plants. I have seen whole rose gardens destroyed by improper use of herbicides.

Use the proper applicator for the job. You will see two full pages of drawings indicating the right equipment used in both home and small commerical jobs. Never spray in a breeze and always know which way the wind is blowing if you are spraying high in the branches. Dormant sprays should never be applied when the plant shows signs of growth.. If you apply a dormant spray to new buds, you will kill them. If you feel you must use poison chemicals to control garden pests, be careful, know exactly what you are doing and be ready when an accident does happen. Have the location and telephone number of your nearest poison control center handy. But I can give you even better advice. Use natural or cultural controls whenever possible.

VICTORIA ASTER.

HAND EQUIPMENT FOR INSECT CONTROL ON VEGETATION

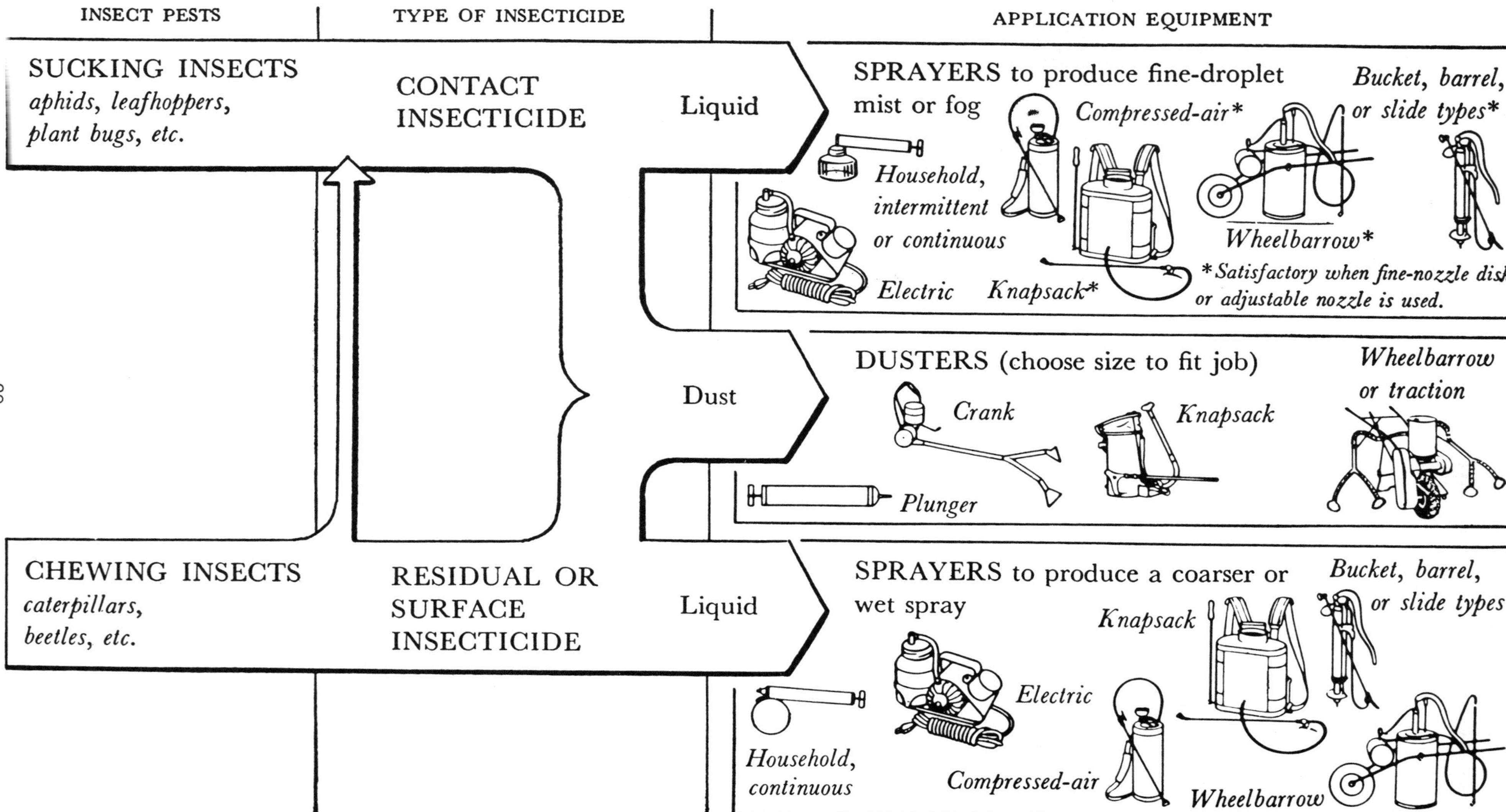

HAND EQUIPMENT FOR INSECT CONTROL IN AND AROUND BUILDINGS

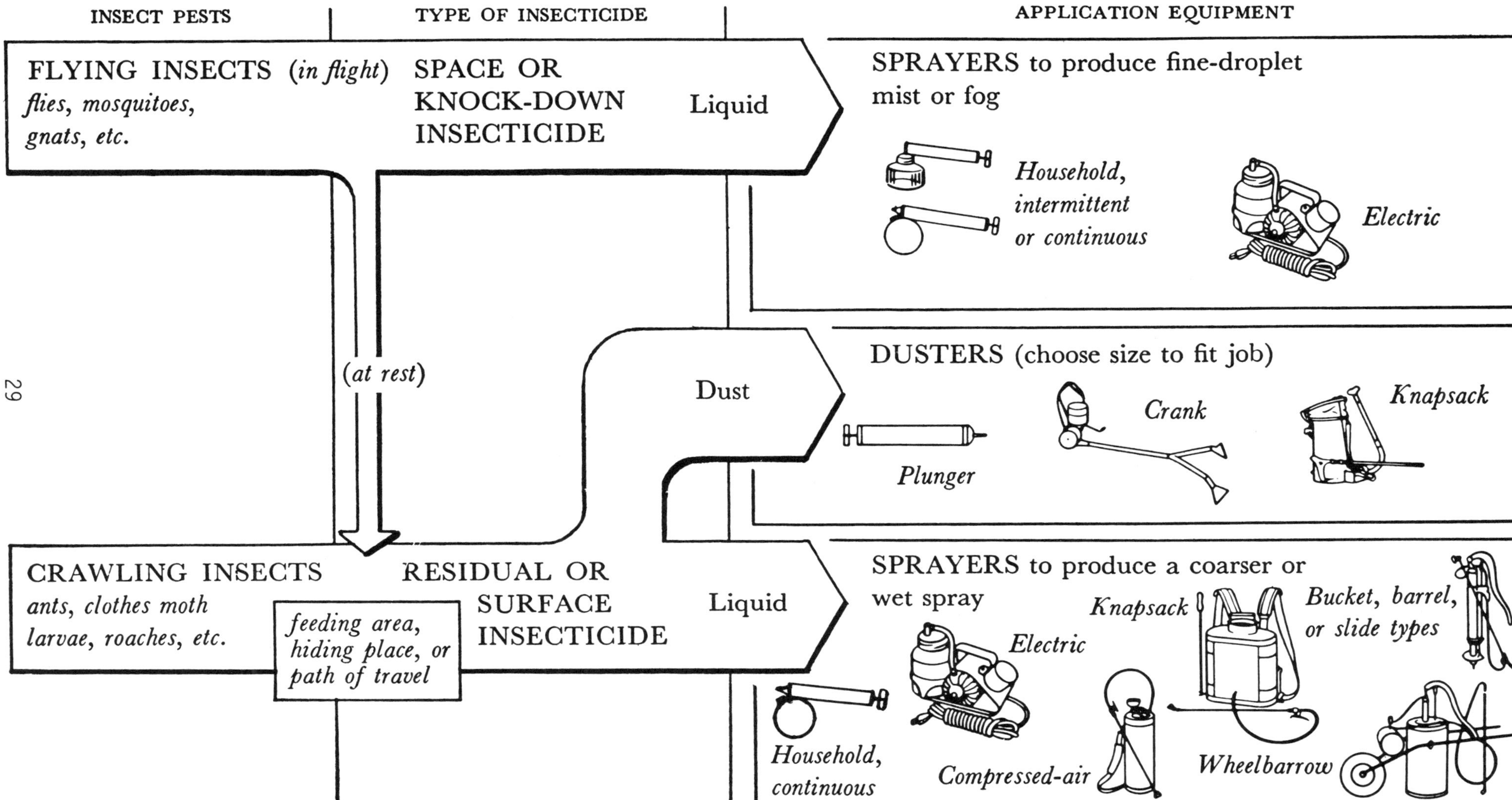

INSECT CONTROL WITHOUT

CHEMICAL PESTICIDES

INSECT CONTROL WITHOUT CHEMICAL PESTICIDES

There are five ways to control insects and mites without recourse to chemicals.

1. Hand control means that you actually pick the insect off the plant and destroy egg masses manually. For this method to be effective you must know what the eggs look like and you'll have no difficultty, of course, finding the larger insects.

2. Mechanical methods include screens over the plants, guards around trees, mechanical traps, light traps, and electrical- traps.

3. Ordinary cultural controls mandate clean garden sites, litter removel, and practices that encourage vigorous growth.

4. Biological controls make use of natural enemies of pests.

5. Purchase of high-quality plants in the first place, and in the case of fruit and nut trees, proper pruning and feeding.

BENEFICIAL INSECTS

Many insects do no damage to plant life but feed on other insects. If you learn to identify these helpful insects and bugs you'll eliminate the need to use a lot of chemicals.

ANT LION

Sometimes called a "doodlebug," this is a half-inch long, brown, rough insect with sickle-shaped jaws. They live in conical pits in the sand, and feed on ants and other pests.
Ant lions are found in most parts of the country.

BENEFICIAL INSECTS

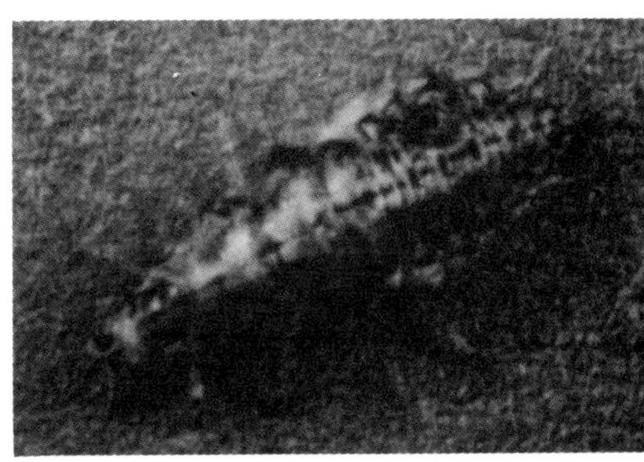

Aphid lion larva

Codlin Moth and Grub (Carpocapsa pomonana).

APHID LION (Lacewing)

The adult is colored soft green and has yellow eyes. Aphid lions have a hair-like antennae and deposit their eggs on the stems of plants. The slender bodies of the larvae are yellowish with red and brown markings. (see illustration) This insect feeds on aphids, mealybugs, thrips, scale insects, and different forms of mites.

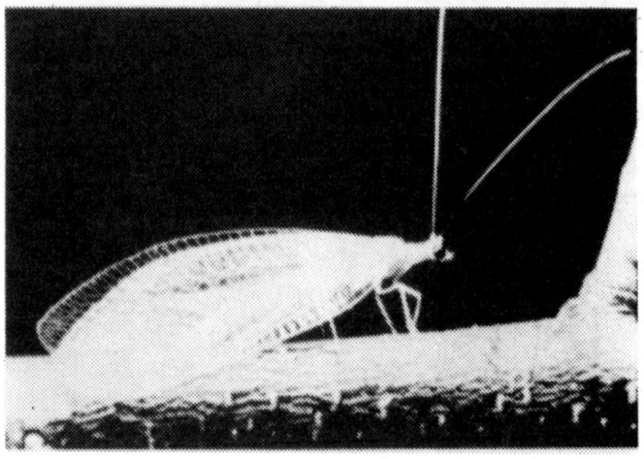

Lacewing

ASSASSIN BUG

A pale brown, slow moving bug with long legs (see drawing), that reaches a length of one half to three quarters of an inch. It holds its prey in forelegs as if it's praying. feeds on immature forms of insects, and is found in most parts of the country.

Assassin Bugs

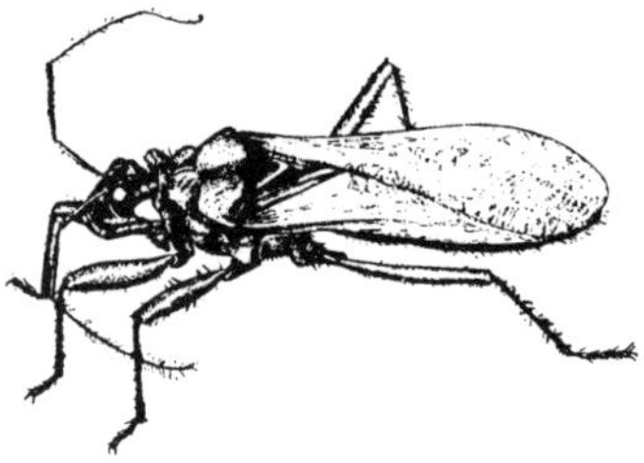

DAMSEL BUG

This light gray bug is like the assassin bug in shape and size, and uses its legs in the same way when feeding. It eats aphids, flea-hoppers, and larvae of most insects. Found in most parts of the country.

GROUND BEETLE

The adult is usually a light black or brown with a long oval body. It hides during the day under trash and feeds at night on cater-pillars and other insects. Found in all parts of the country.

Lady Beetles

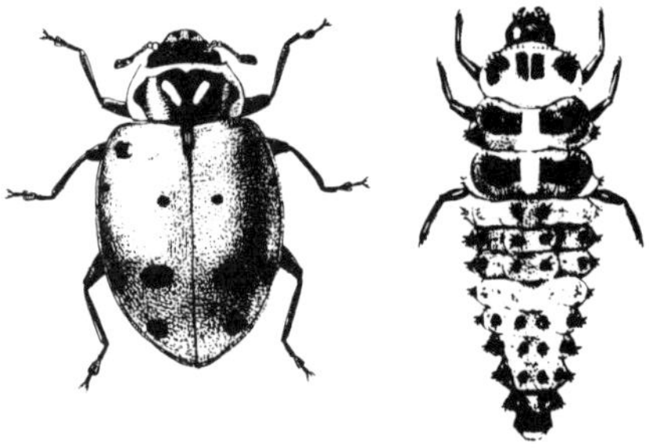

Adult **Larva**

LADY BEETLE

This well-known beetle is associated with organic garden controls. The adult is shiny red or tan with black spots on the back. The larvae is gray or orange and has bumps on the body. It feeds on mealybugs, aphids, scale, and spider mites. Found in all parts of the country.

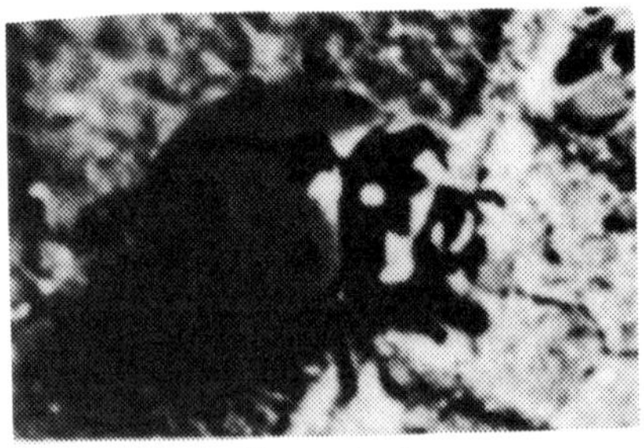

Lady beetle adult

Lady beetle larva

MINUTE PIRATE BUG

The black with white markings adult is flat, and about one sixteenth of an inch long. The nymph is similar but yellowish-tan. Found on the flowering parts of plants and under bark (see drawing), they feed on the adults and eggs of many insects. Found in most parts of the country.

Minute Pirate Bugs

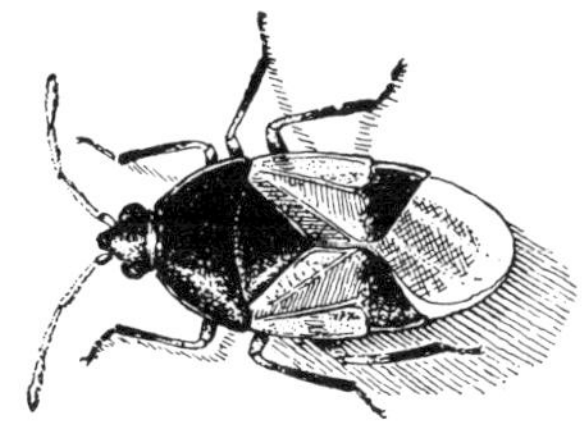

PRAYING MANTID

This strange looking insect (sometimes called a praying mantis) eats all kinds of insects in the garden The green body has green wings with a brown front edge. The large front legs which look as if they were praying, hold the prey. They measure up to five inches. The eggs are laid in grasses and are covered with a froth that then hardens. The young mantids which hatch in the spring, resemble the adults but are wingless. Found in all parts of the country.

Praying Mantids

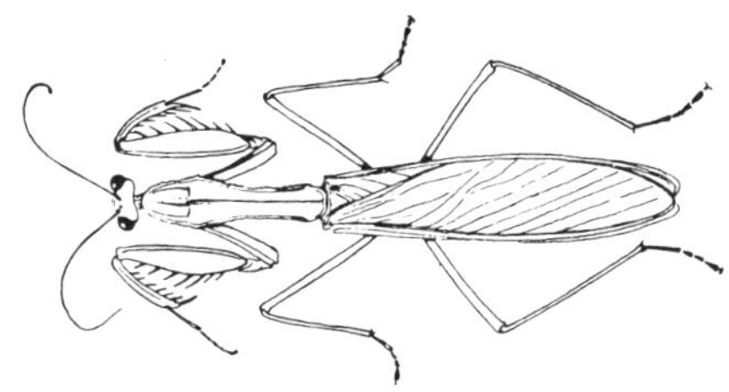

SPIDERS AND MITES

Spiders are excellent for catching many species of insects. The predaceous mites are pinkish or gray. These mites feed on spider mites (red spider), cyclamen mites, aphids, thrips, and eggs of many insects. Both the mites and the spiders are found in all parts of the country.

Syrphid fly adult

SYRPHID FLY

The adult is bright yellow and black and about half an inch long. The larvae resembles a slug and is gray with a mottled coloration. The larvae can eat one aphid per minute, but they do feed on other insects as well. Found in all parts of the country.

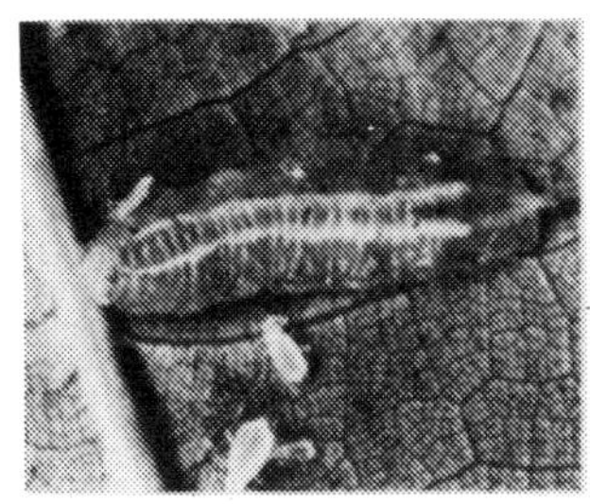

Syrphid fly larva

WASPS

This large family of beneficial insects comes in many sizes and shapes, and are either parasites or predators. The adult wasp lays eggs in the bodies of insects and the developing larvae kill the host. The large predaceous wasp stings caterpillars and feeds them to her young. Found in all parts of the country.

PRAYING MANTID

Some of the developmental stages of a common mantid: a. egg mass attached to stem; b. newly hatched nymphs; c. large nymph; d. adult female feeding on grasshopper.

PLANTS THAT REPEL PESTS

BORAGE (Borage officinalis), Annual, plant seed.
Repels tomato worms

BASIL (Ocimum basilicum), Annual, plant seed.
Repels mosquitoes and flies. Potted, it can be placed near outdoor sitting areas.

ALLIUMS (Perennials), plant bulblets or sets. Garlic (Allium sativum), Onion (Allium cepa). Chive (Allium schoenoprasum), Leek (Allium porrum).
Garlic and onions repel mice and moles and also deter Japanese beetles, aphids. Leeks and chives deter the carrot fly and improve the general health of the plant.

GERANIUM (Pelargonium), perennial, but not hardy in cold climates. Over-winter indoors or propagate from cuttings.
Geraniums (white is most effective) attracts Japanese beetle away from other plants; stuns the beetles which then can be hand picked and destroyed. Plant near rose beds.

PYRETHRUM, the painted daisy (Chyrsanthemum cocineum), perennial.
A source of insecticides, repels aphids.

MARIGOLD (Tagetes), annual, seed.
The old-fashioned smelly varieties provide better control.
Repels nematodes and other garden pests. Use spent plants as mulch around chrysanthemums and clematices. Also repels bean-beetles and Japanese beetles.

MINT (Mentha spicata, and other species), hardy perennials. Once established it spreads.
Deters ants from dwelling and patios, and also deters cabbage moth from laying eggs.

ROSEMARY (Rosemarinum officinalis), tender perennial. Can be potted and taken indoors for winter.
Deters Mexican bean beetle and cabbage moth from chewing foliage.

TANSY (Tanecetum vulgare), perennial.
Repels Japanese beetle, ants, and mosquitoes.

THYME (Thymus vulgaris and other species), perennial. It will spread when established
Deters cabbage worm.

GERANIUM—"MRS. POLLOCK."

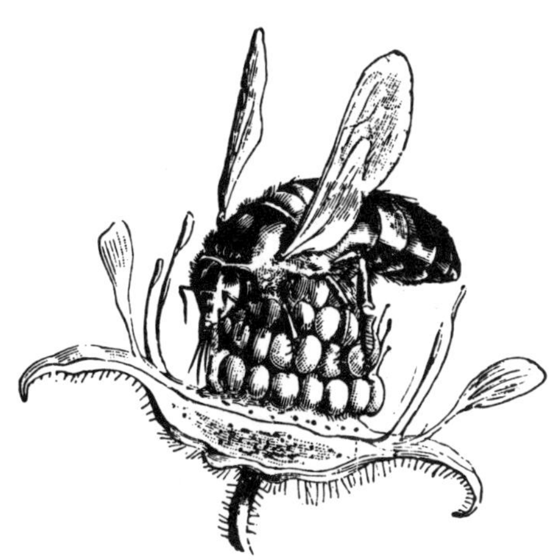

REPELLENTS FOR VEGETABLES

ASPARAGUS-Plant near tomato.
BEANS-Green beans-plant near potatoes, Lima-nasturtium, garlic, savory, Pole-marigolds, savory, Peas-compost in furrow, wood ashes, Snap-garlic.
BROCCOLI- sage, mint.
CABBAGE- mustard, sage, tar paper squares, mint.
CAULIFLOWER-sage, mint.
CUCUMBER- radish, nasturtium.
EGGPLANT- mulch, basil, hot pepper.
KOHLRABI- marigold, sage, mint.
LETTUCE- garlic, mulch.
MELON- radish, nasturtium, brew made with garlic, onion, hot pepper, peppercorns, etc., mixed with one cup detergent to one gallon water. Let stand and brew to one quart

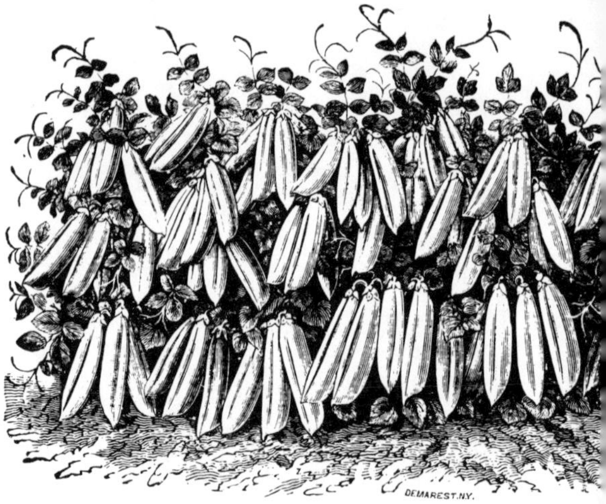

—AMERICAN WONDER PEA.

MELON-radish, nasturtium, brew made with garlic, onion, hot pepper, peppercorns, etc., mixed with one cup detergent to one gallon water. Let stand and brew. Use one half cup of this brew to one quart water and douse plants.
SEEDLINGS-same brew as above.
TOMATOES-basil, dill, plant near asparagus.
TURNIP- tar paper squares.

OTHER USEFUL FLOWERS AND HERBS-asters, anise, artemisias, dahlias, tarragon, yarrow, and hyssop.

—EARLY ROUND DARK RED RADISH.

ORGANIC INSECTICIDES

The following is a list of **substances frequently used or recommended by organic gardeners, although purists do not rely on any pesticides.** Instead they pay meticulous attention to horticultural and non-chemical control methods, taking pains not to disturb their crop's environment.

NICOTINE

Nicotine is an alkaloid derived from plants in genus Nicotiana and is well known as a greenhouse insecticide. It is usually found as a 40 percent alkaloid formulation; used as a fumigant, dust or spray, and can be used on many crops and ornamentals.

Pests Controlled. Aphids, thrips, and mealybugs.

Environmental Impact. It is highly to moderately toxic to man, especially if inhaled; has a rapid breakdown (little sidual effect), is not harmful to bees 1 most beneficial insects.

RYANIA

Ryania is a botanical derived from Ryania speciosa roots and stems, is a contact and stomach poison, and is not commonly found.

Pests Controlled. Cranberry fruit worms, European corn borers and caterpillars (codling moth) on apple.

Environmental Impact. Slightly toxic to mammals.

PYRETHRUM

Pyrethrum is a botanical derived from certain dried flower heads in the genus chrysanthemum and comes in dust or liquid formulation. It is a contact insecticide with rapid knockdown that affects the nervous system of the insect. It can be used on most fruits and vegetables.

Pests Controlled. Ants, aphids, asparagus beetles, cabbage loopers, cheese-mites, crickets, fleabeetles, flies, fruit flies, gnats, horseflies, leafhoppers, leafrollers, psyllids, roaches, spiders, sodwebworm, wasps and ticks.

Environmental Impact. It is slightly toxic to man, not hazardous to birds and other wildlife with extremly low toxicity to mammals. It is nontoxic when used on pets (dogs, cats, birds),and can be used in food areas. There is a broad spectrum of insecticide activity, and lack of persistence and biomagnification in the food chain. Because it irritates insects, it acts as a flushing agent, making them leave their hiding places. There is a rapid breakdown when exposed to sunlight and alkali.

ROTENONE

Rotenone is a non-phytotoxic botanical derived from derris or cuberoots. It is primarily a contact insecticide but also acts as a stomach poison and inactivates an enzyme system causing death from lack of oxygen. It is registered for most fruits and vegetables.

Pests Controlled. Aphids, cabbageworms, Colorado potato beetles, asparagus beetles, ants, cucumber beetles, European corn borer, bean beetles, lice, greenhouse white flies, pepper maggots, rose chafers, and squash bugs. It also kills spiders and mites.

Environmental Impact. It is moderately toxic to man, extremely toxic to fish and other cold blooded animals, and is reported highly toxic to pigs. The hazard to birds and wildlife is low.

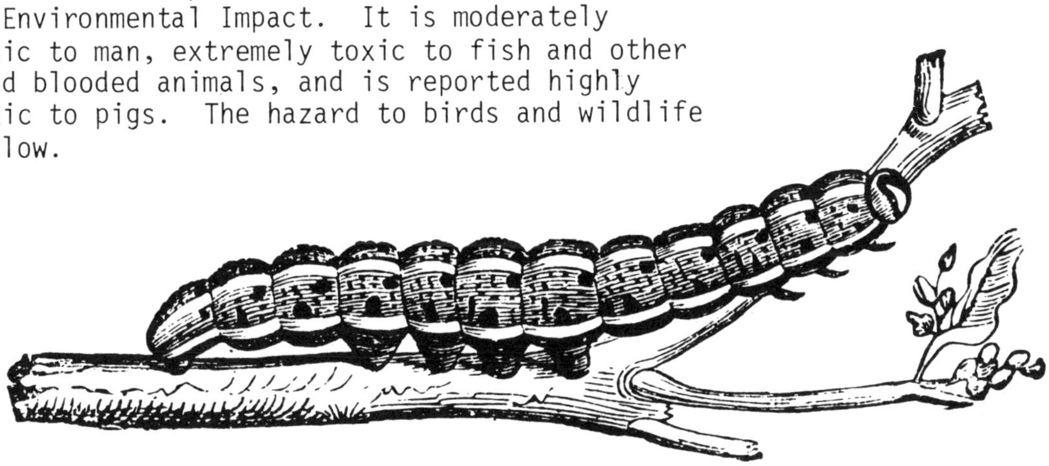

MILKY SPORE DISEASE

(Natural control of Japanese beetle)

The Japanese beetle attacks over 200 kinds of food and ornamental plants, ranging from flowering trees and shrubs to vegetables of all kinds to most garden flowers. The damage runs into the millions each year for commercial growers and causes problems for the home gardener.

Two kinds of bacteria cause milky spore in Japanese beetle grubs, and forty other species of white grubs. It is important to remember that this is a natural control and does not harm birds, animals, or humans. It is also totally harmless to plant life.

The spores of the milky disease rest in the soil and in the bodies of the grubs that have been affected by the bacteria. The spores are too tiny to really see, but you can see the effect when the white grubs turn a milky color. The milky spore dust is applied to the soil and will last for many years, killing the grubs for many generations. The spores are resistant to dry conditions, excessive moisture, and most adverse weather conditions. They seem to be the perfect method of controlling the beetle in the grub stage. A mixture of chalk dust or talc with the spores provides a protection from ultraviolet sunlight and renders the spores more potent.

HOW IT WORKS

As the grub tunnels through the soil in search of roots and other food, the spores which are carried in the soil, are digested by the grub as it eats the roots and organic matter. Once in the grub, they germinate and multiply in the grub's blood. When the bacteria become very numerous, they again form spores, thus completing the cycle.

Many chemical insecticides kill the insect but natural controls, even though they take a little longer to work, last longer and will not damage the environment. A lot of communities are working together to spread the milky spore control. The local application of milky spore dust will not prevent the beetle from leaving the area and infesting the next-door garden. It is a long range control.

The availability of milky spore is widespread and can be found in most garden shops or landscape centers. You can ask your local United States Agriculture Extension agent for a listing of

Bush.

where to buy the spore in your area.

The spores will not remain in the soil if the Japanese beetle grub is not present, as they totally depend on the grub to complete their life cycle. The more grubs in the soil, the better the spore will take hold. Many experts say the control is effective only on established lawns or permanent bed sites. If the soil is cultivated, the spores will die. You can apply the dust anytime during the fall or spring, except when the turf is frozen.

The best method of application (see photograph) is to place the dust in small piles in a ten by ten pattern. Apply one level teaspoon at each spot. This is a rate of two pounds per acre. You can speed up the results by spreading the dust with a spreader, but never use more than twenty pounds per acre.

Many factors are involved for the effective establishment of the milky spore disease. Besides the need for grubs in the soil, the soil must be between 60 and 90 degrees. In parts of the country where the temperature is above 70 degrees for a long period of time, the disease will develop very fast and build up in the adjoining areas as well.

The federal government, in cooperation with state agencies spread over 229,000 pounds of this dust during the period of 1939-1953. This greatly reduced the establishment of the beetle in many parts of the country and prevented major damage. Research is continuing to determine why the beetle still develops in regions where the dust has been spread (even though out-breaks are very localized). Temperature factors may have something to do with this problem. The reduction in Japanese beetle population has been vast but more work should be done in the field. We can only hope the government spends more on agriculture research and less on paperwork.

TOBACCO HORNWORM AND AN ENEMY

A, Parasite cocoons and adult on hornworm (greatly magnified); *a*, hornworm eggs; *b*, young larva; *c*, mature larva; *d*, pupa; *e*, adult moth. (All stages of hornworm about natural size.)

Applying milky-disease spore dust. Here a hand corn planter is used.

Dead turf rolled back, exposing beetle grubs that have been feeding on the roots.

The spring Tiphia, a wasp-like insect parasite that lives on and destroys grubs of the Japanese beetle.

Grubs: A, normal: B: diseased

VEGETABLE PESTS

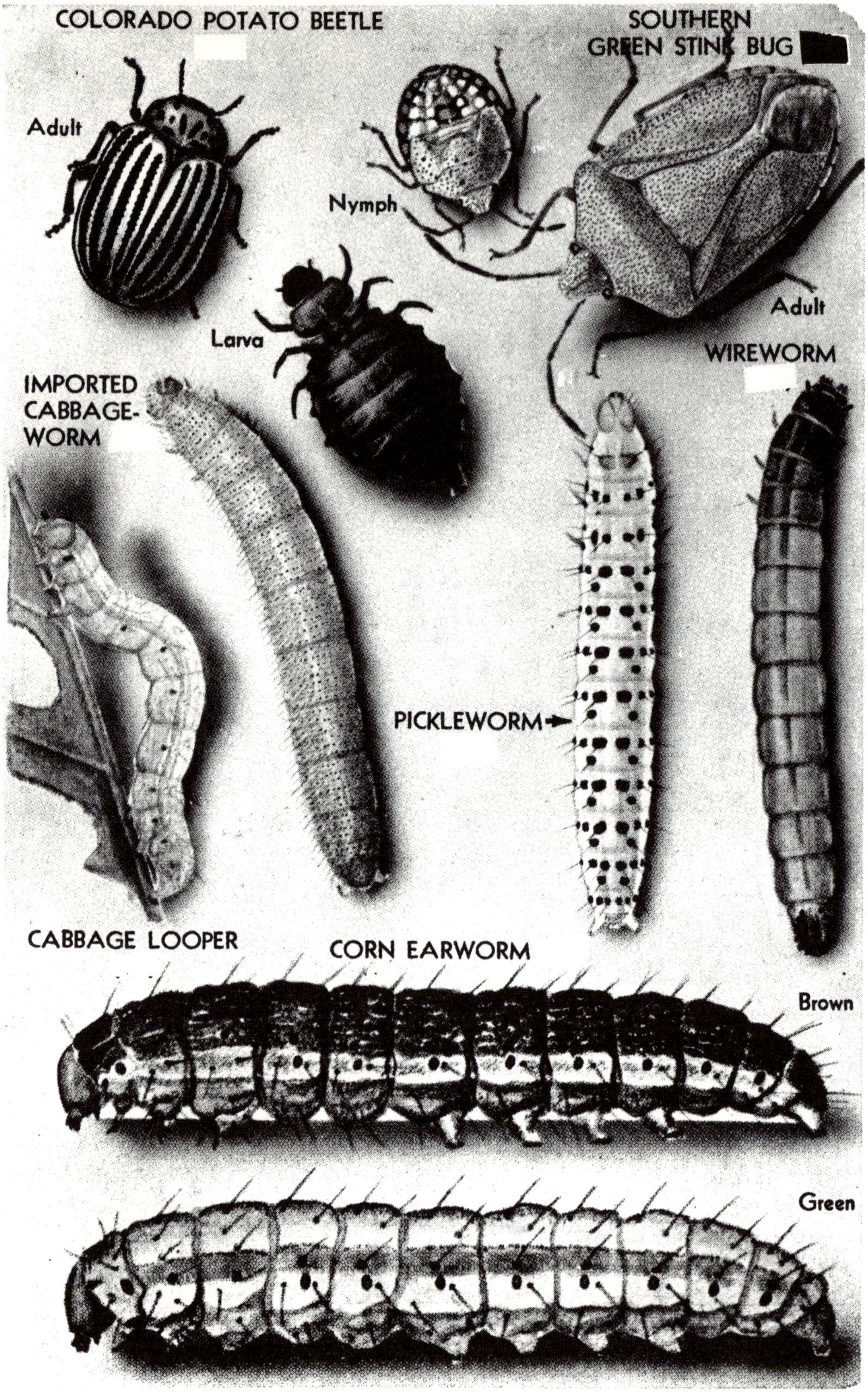

VEGETABLE PESTS

PESTS OF ASPARAGUS

ASPARAGUS BEETLE

Adult: Metallic blue to black. Orange and yellow markings. About one quarter of an inch long.
Larva: Olive green to dark gray. About one third of an inch long. The eggs look like black specks.
Damage: Adults and larvae eat foliage.
Location: Generally found in all parts of the country.

RUST (FUNGUS)

This is a very dangerous disease, and the only real threat to asparagus. Its symptoms are orange-red powdery blisters on the stem and foliage. Rust increases in the moist weather of the growing season and lives on the remains of the plant after the season is over. Grow rust resistant varieties such as Mary Washington and Martha Washington. Destroy all diseased tops.
Location: All sections of the United States and Canada.

PESTS OF BEANS

BEAN APHID (PLANT LOUSE)

The bean aphid looks like tiny black specks on the underpart of the leaves and on the stems. Both the young and the adult appear the same.
Damage: The leaves curl and turn yellow. This aphid carries the common bean mosiac disease. (In the section called "Insects and Diseases", all the insects that are disease carriers are listed). It is very important to understand the natural carriers of plant disease in order to control them without an excessive use of fungicides.
Location: Found in all parts of the country.

BEAN LEAF-BEETLE

Adult: A red-yellowish color with black spots on the top, about one quarter of an inch long.
Damage: Eats foliage
Location: Eastern United States.

CORN EARWORM

This insect can be either pink, brown, or green, with a stripe down one side. It is about three quarters of an inch long. (On tomatoes it is called a fruitworm.)
Damage: Eats holes in the bean pod and then attacks mature beans in the late summer and fall.
Location: Most parts of the country.

CUTWORM (See Tomato)

LEAFHOPPER

There are several different kinds. The adults are green,wedge shaped, and less than one quarter of an inch long. The leafhopper will jump or fly if the plant is shaken. The nymphs look like the adult but crawl sideways like a crab.
Damage: Both adults and nymphs attack beans. The foliage will turn yellow or brown and often the plant will die.
Location: Throughout the United States and Canada.

LIMA BEAN POD BORER

This insect is about one inch long and has a yellow head and a pinkish body. If disturbed on the plant it will make very fast movements.
Damage: Bores into the pod and eats the beans. Usually does not trouble snap beans.
Location: Southern United States and California.

MEXICAN BEAN BEETLE

A. Leaves skeletonized by bean beetle feeding. a. egg cluster; b. larva; c. pupa with larval skin still attached; d. adult. (all greatly enlarged).

CORN EARWORM

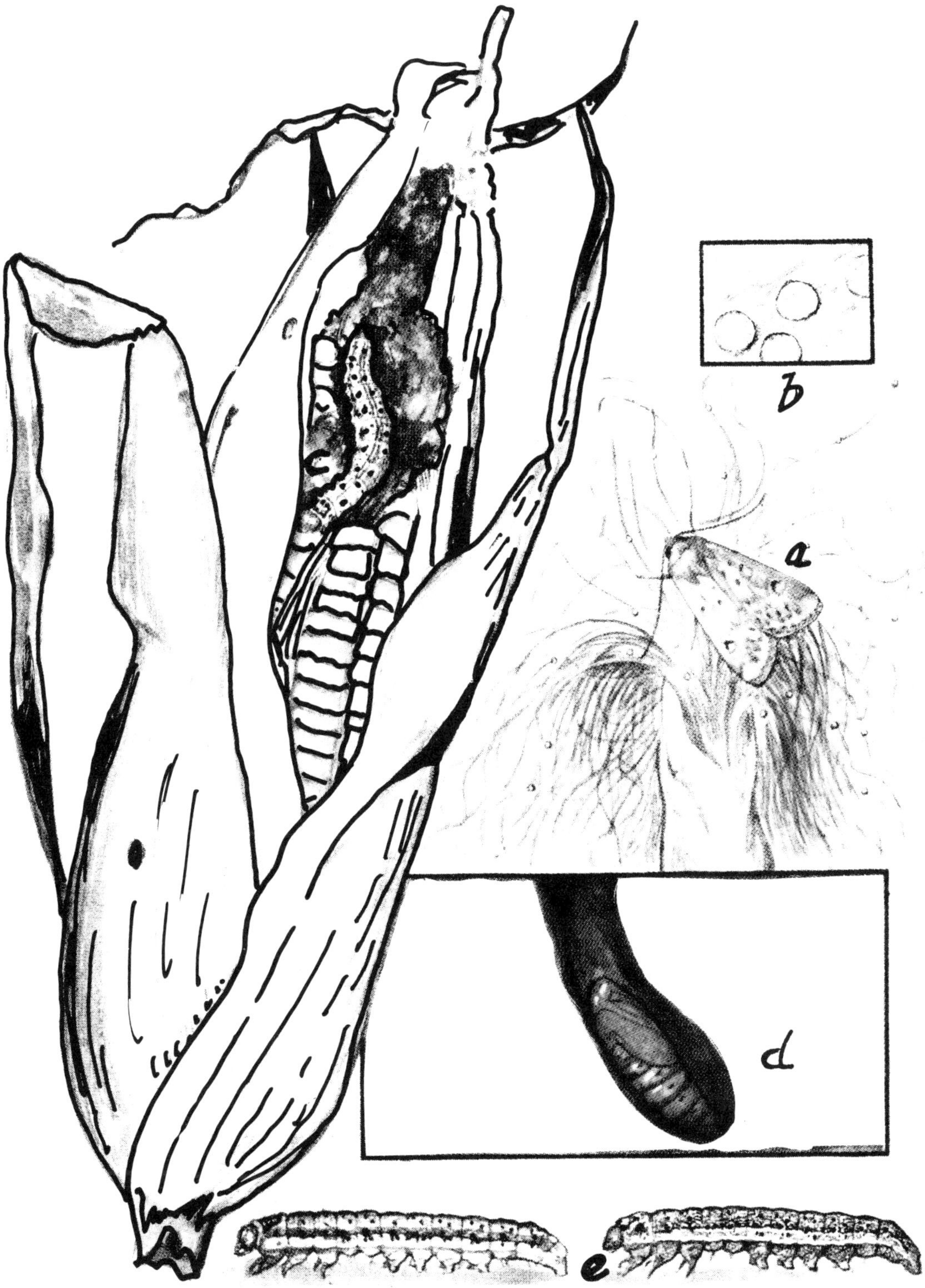

a. Moth (or adult) and eggs on silks; b. eggs: c. earworm feeding in ear of corn; d. pupa in a cell; e. coloration of the earworm; (all sizes larger than natural size.)

LYGUS BUG

There are several species but the tarnished bug is by far the most common and does the most damage. The bug is flat, black and white, and about one quarter of an inch long. You have to look closely for this bug, for if moved it will jump to another plant. Damage: Both the adult and the young suck the juices from the foliage and the bean pods, causing the pods (and sometimes the leaves) to drop off. On the older plants the pods become disfigured and the beans in the pod become pitted and undesirable for eating.
Location: Throughout the United States and Canada.

MEXICAN BEAN BEETLE

Has a copper color with sixteen black spots on the back and is just over one quarter of an inch. The larva is yellowish-orange and fuzzy.
Damage: These beetles usually feed on the underside of the leaves and strip the foliage off the plant, they also attack the pods.
Location: East of the Rocky Mountains. Very seldom found on the West Coast.

SEED-CORN MAGGOT

This is a large, light yellow maggot about one quarter of an inch long.
Damage: Most of the damage is done to the young transplanted bean sprout. It bores into the seed and destroys the plant before it has a chance to develop. Use insect-free soil in the transplanting beds, turn the soil over well and look for signs of this maggot before planting the bean seeds.
Location: Found throughout the United States and Canada.

SPIDER MITES (See Muskmelons)

SPOTTED-CUCUMBER BEETLE

Light yellow and green with twelve black spots on the top and about one quarter of an inch long.
Damage: This chewing insect eats holes in the foliage and chews on the pods.
Location: East of the Rocky Mountains but related species are found on the West Coast and in almost all sections of the country and Canada.

ANTHRACNOSE (FUNGUS)

The pods become brown with pink centers the stems develop dark red cankers, and reddish colors develop on the veins of the leaves. The seeds have brown spots. This disease spreads most rapic during moist, cool, summers. The fungu is carried both on the seed and other parts of the plant.
Location: Southern States, Central States, and parts of the Northeast. Do not plant seeds that are discolored or from spotted pods. If possible, grow beans that have been propagated in the Western part of the country. Never work the bean rows when it is wet, for the disease spreads more rapic at this time.

BACTERIAL BLIGHTS

The foliage dries out and brown spots appear on the leaves, the pods have water-soaked spots with red margins, and red cankers develop on the stems. This disease is carried on the seed.
Location: Not found west of the Rockie so plant Western seed if possible and follow directions mentioned in Anthracnose.

BEAN MOSAIC (VIRUS)

The plants become stunted and the foliage turns light and dark green. Many times the leaves curl. This virus is carried on the seed and spreads by aphids. You can grow mosaic resistant varieties such as:

Wade
Wax
Tendercrop
Kentucky
Wonder(pole)
Blue Lake (pole)

RUST

The foliage blisters and turns red and black. As the disease progresses, the leaves turn completely dark yellow and drop from the plant. Rust lives on old plants from the last growing season.

A. Normal alfalfa plant; a. undamaged bloom; b. well developed seed pods;
c. nymphs; d. adult lygus bugs;
B. Damaged plant

SEED-CORN MAGGOT

(on lima bean)

A. Life stages; a. eggs; b. larva; c. pupa; d. adult;
B. Injury to plant seedling C. injury to young plant
D. Young plant with part of stem cut open to expose maggots inside.

SEED DECAY (FUNGUS)

Seeds are rotted. This is a common disease found in the soil. It can't be controlled except by buying treated seeds from a reputable seed house.
Location: All parts of the country.

Bean Mosaic

Beet Webworm

PESTS OF BEETS AND CHARD

BEET WEBWORM

A greenish-yellow worm with black spots and a black stripe, it is about one and a quarter inches long.
Damage: Eats the leaves and the new buds. It folds the leaves and ties them together with a webbing material.
Location: Found in all parts of the country.

BLISTER BEETLES

Many species--some are black, some have stripes, all are about one half an inch long.
Damage: Eats foliage.
Location: Found in most parts of the country. Not a pest until late in the season. If you handpick these beetles be careful and wear gloves for they discharge a caustic fluid that may blister your skin.

FLEA BEETLE

This is a very small jumping beetle and comes in many colors. Many species are stripped, some are brown, others black.
Damage: Eats small holes in the foliage of all varieties of vegetables.
Location: Found in all parts of the country and Canada.

Blister Beetles

DAMPING-OFF (FUNGUS)

The seed decays in the soil, or the young plants wilt and die. You can buy treated seeds that will prevent this soil fungus. If you prefer to use your own seeds, use soil that has been sterilized.
Location: Very common and found everywhere.

LEAF-SPOT (FUNGUS)

Many spots on the leaves with dark brown margins and light brown centers. This fungus is carried on the seed and over-winters on old plants. If you keep the garden free of all old plant parts many diseases will be prevented.

BROCCOLI (See Cabbage)

BRUSSEL SPROUTS (See Cabbage)

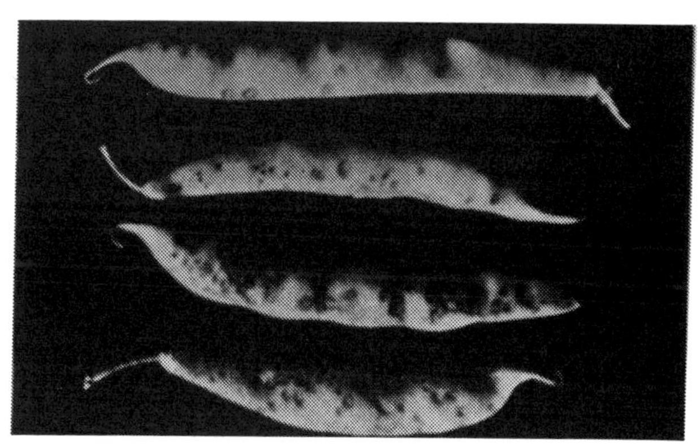

Bacterial blight of bean

IMPORTED CABBAGEWORM

a. butterflies (or adults) with wings in natural positions; b. larvae (caterpillars, or worms); c. pupae (in the resting stage); d. cabbage plant showing typical feeding injuries. (all enlarged).

PESTS OF THE CABBAGE FAMILY

(BRUSSEL SPROUTS, BROCCOLI, CAULIFLOWER)

CABBAGE APHID

A very small greenish-blue insect with a fine whitish wax covering over its body. As with all aphids, cabbage aphids cluster on the foliage. Try to remove all damaged plant parts early in the season. Also try companion planting to discourage aphids. Check the section on natural and chemical controls of plant pests.
Damage: Leaves tend to curl and plants become stunted. Damages kale, brussel sprouts, collards and broccoli.
Location: While it is found in all parts of the country, it is a real problem in the South.

CABBAGE LOOPER

A light green measuring worm with light stripes down the back. About one and a half inches long. As it "crawls" it loops or doubles up.
Damage: Eats the foliage and burrows into the cabbage head.
Location: Continental United States.

CUTWORM (See Lettuce)

DIAMONDBACK MOTH CATERPILLAR

The larva, a light green caterpillar about one third of a inch long, is the pest. If disturbed it hangs from a thread.
Location: Throughout the United States and Canada.

IMPORTED CABBAGEWORM

Light green with a yellow stripe on the back. It is about one and a quarter inches long.
Damage: Eats foliage and bores into the cabbage head.
Location: Throughout the United States and Canada.

HARLEQUIN BUG

A brightly colored bug with red and yellow markings. It is just over half an inch long in the complete adult form.

The white butterfly that lays the eggs for the cabbage-worm.

Damage: Eats foliage so that the leaves turn brown and the plant wilts.
Location: Southern United States and California.

ROOT MAGGOTS

Light yellow to white and about one quarter of an inch long. Many species, but the cabbage and corn maggot are the most common.
Damage: Very destructive in the seedbeds and does a lot of damage to transplants. It tunnels in the roots and lower stems of the plant which wilts and kills the young plant.
Location: The seed corn maggot is found throughout the United States but the cabbage maggot is found in the Northern regions.

VEGETABLE WEEVIL (See Turnips)

BLACKLEG (FUNGUS)

Gray spots and small black speckled dots on the stems and leaves. The plants wilt and die. This fungus is carried on the seed and lives on old plant parts.
Location: Most parts of the country.

HARLEQUIN BUG

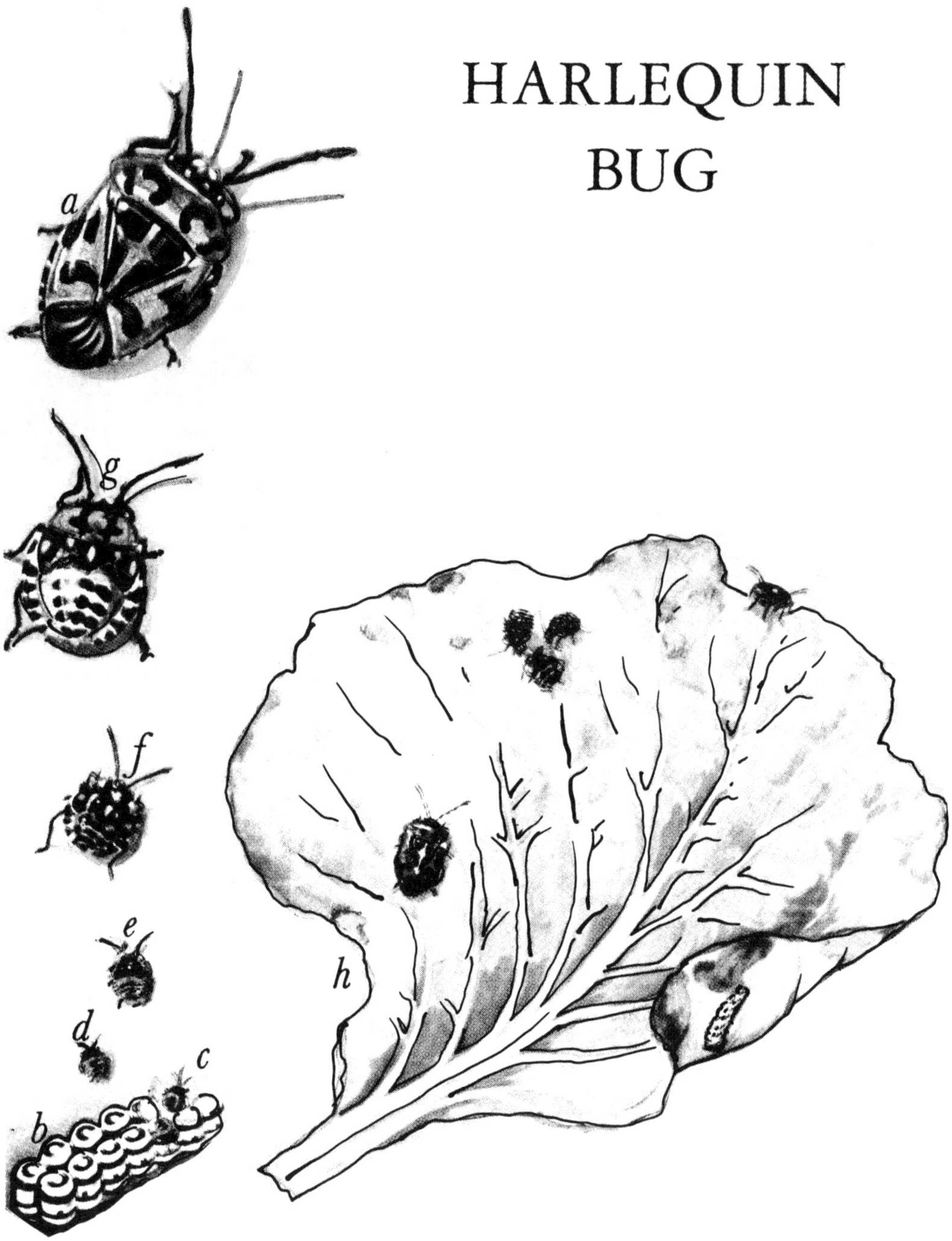

a. adult; b. eggs; c. young, or nymphs (d, e, f); h. damaged cabbage leaf with nymphs, adult and eggs. (all greatly enlarged).

BLACK ROT (BACTERIAL)

The stems turn black and the leaves turn a pale yellow and drop from the plant. In severe outbreaks the plant may die. This bacteria lives in the soil and is transmitted on the seed.
Location: Found in the Central, Southern, and Eastern States.

CLUBROOT (SLIME MOLD)

Large clubs form on the roots and the plant becomes stunted. If this infestation gets into the seed beds it will cause a lot of damage. The mold enters the roots through the soil.
Location: Found in all parts of the country.

DAMPING-OFF (FUNGUS)

The seeds decay in the soil and the tender new seedlings are killed.
Location: Found in all parts of the country.

PINK ROT (See Celery)

RHIZACTONIA DISEASE (FUNGUS)

Sometimes called wire-stem. This disease turns the stems dark just above the soil line. Older leaves turn dark and droop but do not drop. The cabbage head may rot.
Location: This fungus lives in the soil in most parts of the country.

WILTS OR YELLOWS (FUNGUS)

Lower leaves drop off the plant and the foliage turns a yellowish-green. The plants become stunted. The plants this fungus attack include cabbage, kale, and kohlrabi. Resistant members of the family are cauliflower, broccoli, and brussel sprouts.
This disease usually attacks only one side of the plant. The fungus lives in the soil and enters the plant through the root systems. You can grow disease resistant varieties such as:
Jersey Queen, Marion Market, Globe, and many more that are listed in vegetable seed catalogues. It does not help to spray the plants.

PESTS OF CARROTS

Carrot Caterpillar

CARROT CATERPILLAR

Green with black and yellow markings. About two inches long.
Damage: Eats the foliage and the tops. They seldom come in large numbers, so you can control them by hand picking.
Location: Found in all parts of the country.

Carrot Rust Fly

CARROT RUST FLY

The light yellow to white larva is the pest. About one third of an inch long.
Damage: Tunnels into the roots and destroys the fibrous root systems of the carrot.
Location: Pacific Northwest and Northeast.

SIX-SPOTTED LEAFHOPPER (See Lettuce)

VEGETABLE WEEVIL (See Potato)

WIREWORM (See Potato)

LEAF BLIGHT (FUNGUS)

Black spots appear on the stems and foliage. Older leaves drop off and turn a brownish-yellow. The fungus is carried on the seed and the remains of last year's crop. Found in all parts of the country. It is advisable to rotate vegetables to different parts of the plot to prevent this from happening the next growing season.

YELLOWS (VIRUS)

The foliage becomes yellow at the center of the crown, the tops follow the yellowing pattern, and old leaves become red and twisted. This is the same virus that causes aster yellows and it attacks many cultivated and wild plants. It is spread by the leafhopper that lives on perennial plants, so control of this insect will help prevent the disease.
Location: Found in all sections of the country.

PACIFIC COAST WIREWORM

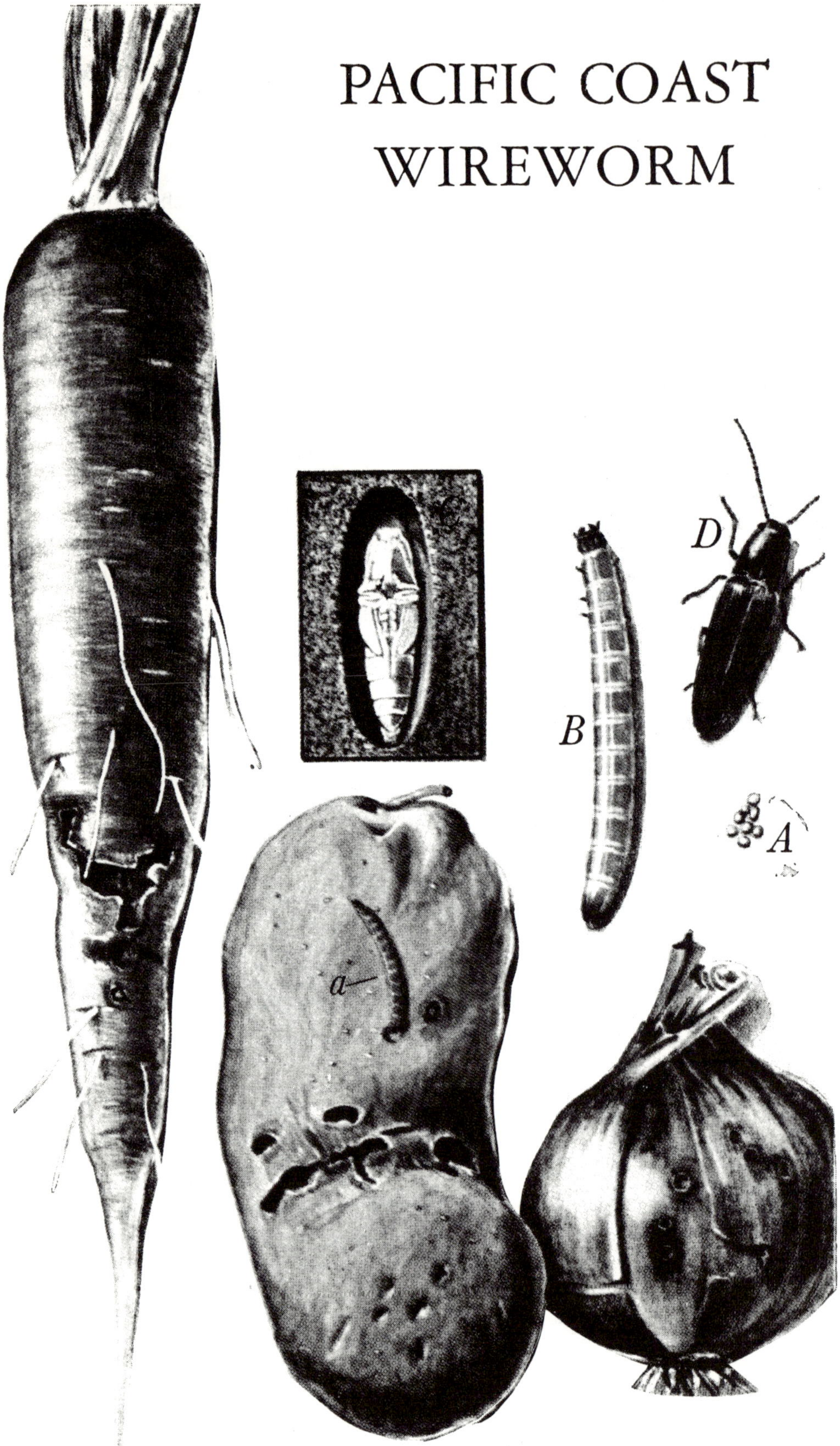

Damage to carrot, potato, and onion; a. larva on potato;
A. eggs; B. larva; C. pupa in underground cell; D. adult click beetle.

CAULIFLOWER (See Cabbage)

PESTS OF CELERY

APHIDS (See Muskmelon)

CELERY LEAF TIER

Green and about three quarters of an inch long.
Damage: Eats foliage and stems. It folds the leaves and binds them together with a web.
Location: Found in all parts of the country.

DAMPING-OFF

Seeds decay in the soil and young seedlings are killed. To control, buy treated seeds and propagate all plants in clean seed trays with a sterilized soil or a soil mix sold just for propagating seeds.

EARLY BLIGHT (FUNGUS)

Small, round yellow-brown spots appear on the older leaves. As the spots get larger they turn light gray. Early blight attacks the celery stalks and is spread by the rain. This fungus lives in the soil.

LATE-BLIGHT (FUNGUS)

First signs are yellow spots on the stalks and foliage. As the disease develops, tiny black dots appear on the stalks and they turn gray.
Location: It is found in all parts of the country.

PINK ROT (FUNGUS)

You will see a pink-white fuzzy growth at the base of the stalks which become water soaked. The celery stalks rot and have a bitter taste. This fungus also damages lettuce and cabbage. It remains in the soil and may remain there for many years. Pink-rot is found in the Northeast, North Central and Southern States. Try rotation of celery, lettuce, and cabbage plants to different parts of the vegetable plot. Destroy all old plants.

Celery Leaf Tier

YELLOWS (FUNGUS)

Foliage turns yellow and the plants become stunted. You can grow disease resistant varieties if you like the self-blanching kind. Some of the better varieties are:
Forbes Golden Plume
Michigan Golden
most of the green celeries as they are disease resistant.

CHARD (See Beet)

COLLARDS (See Cabbage)

CORN (See Sweet Corn)

PESTS OF CUCUMBER

APHIDS (See Muskmelon)

PICKLE WORM

The adult is light yellow with a light brown head. About three quarters of an inch long. Young worms have some dark spots on their bodies.
Damage: Eats the flowers and buds. Tunnels in the terminal shoots of the vine.
Location: Found in the southeastern part of the country but has been seen in some northern regions. Over-winters in the South and spreads North as the warmer weather approaches.

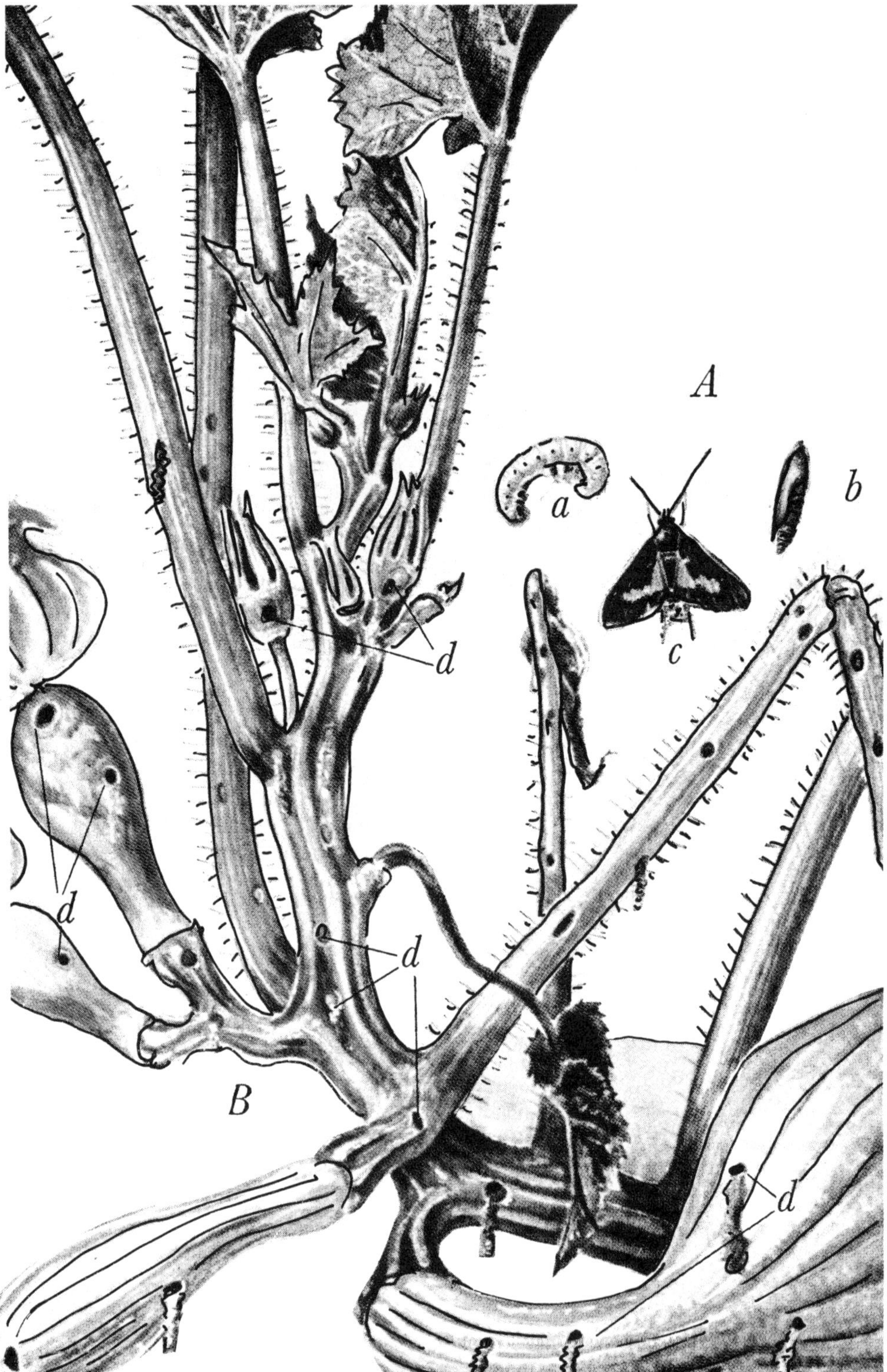

Pickleworm

A. Life stages: a. larva; b. pupa; and c. adult.
B. A small part of the vine showing damage.
d. feeding holes in flower buds and stem (all enlarged).

STRIPED CUCUMBER BEETLE

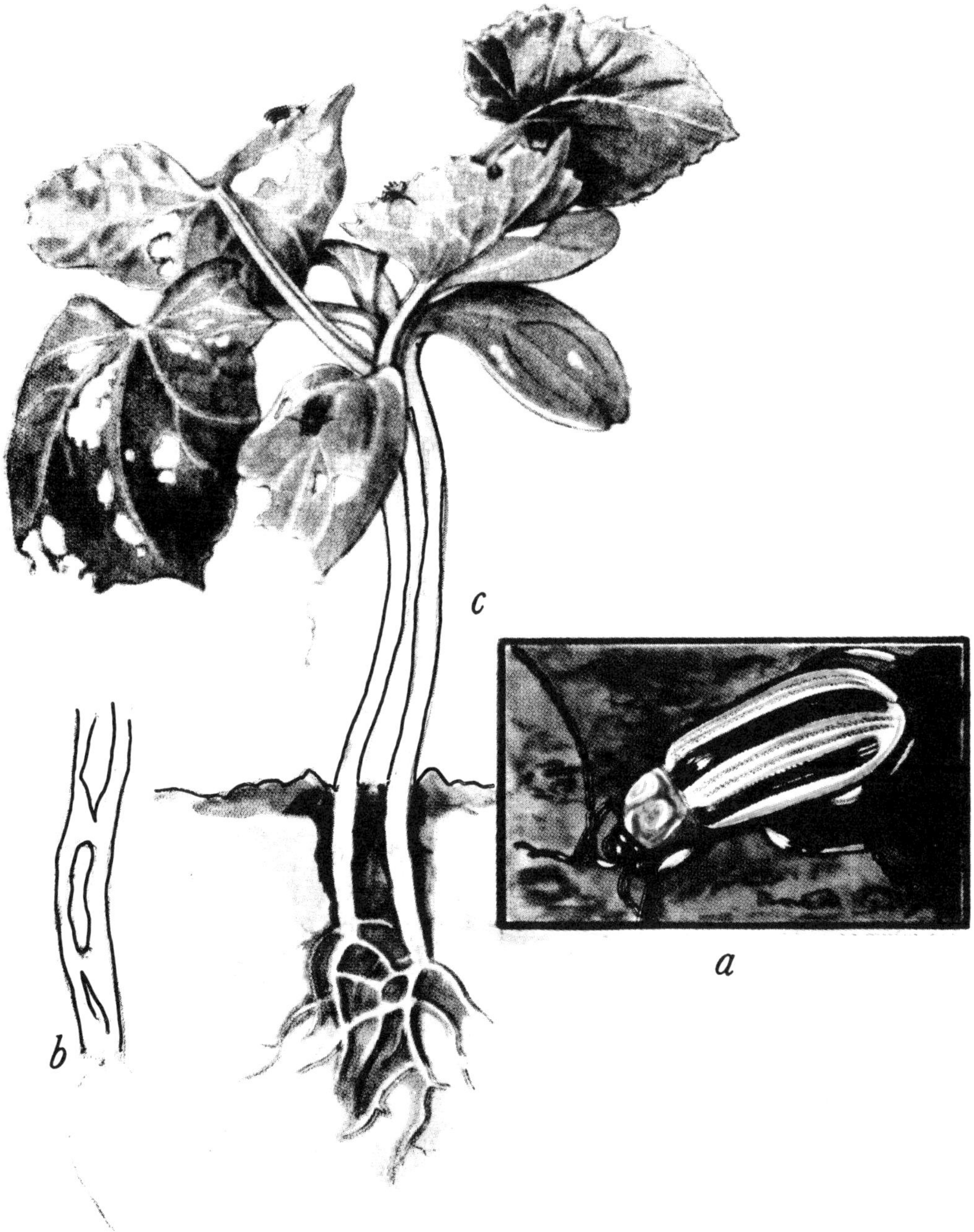

Striped cucumber beetle

a. adult beetle; b. underground stem of cucumber seedling cut open to show larva (grub or worm) feeding: c. small cucumber plants showing feeding habits by adult beetles on leaves and stems; (all enlarged).

STRIPED CUCUMBER BEETLE

Yellowish-black with three black stripes down the top. About one fifth of an inch long. The larvae is white with brown tips at the ends. Very slender. Location: Mostly East of the Rocky Mountains, but some related species are found on the West Coast.

SPIDER MITES (See Muskmelon)

ANTHRACNOSE

Brownish-red spots develop on the foliage and elongated brown cankers on the stems. The cucumbers develop sunken pinkish spots that turn dark reddish-brown. Most damage is done during the moist seasons. This fungus, like most fungi, is carried on the seed and lives in the soil. All melon crops should be moved every three years to another part of the garden. Location: Found in the Central, Southern, and Eastern States.

BACTERIAL WILT

The large leaves wilt and die (there is no yellowing). When the plants are very young they die rapidly. This bacterial disease is transmitted by the cucumber beetle. It enters the water vessels of the plant and plugs the passages. Bacterial Wilt is found in the Northern, Central and North-eastern States. As soon as this disease is detected, remove the plants from the garden and destroy. No cure.

DAMPING-OFF (FUNGUS)

Seeds decay in the ground and the young seedlings wilt and die. As in all cases of this common fungus, the seeds should be treated and grown in clean soil.

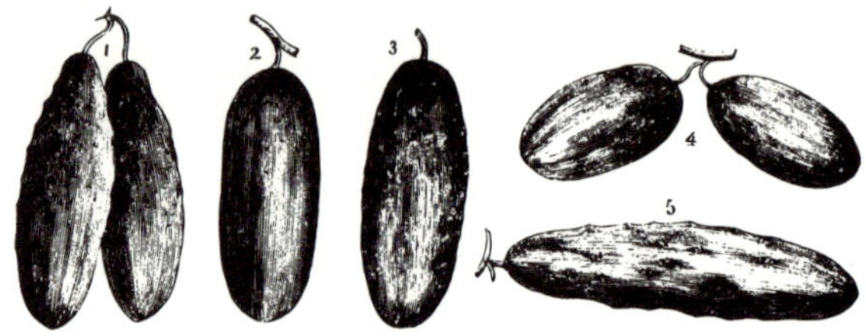

DOWNY-MILDEW (FUNGUS)

The older leaves turn yellow with somewhat angular spots, then curl up, dry out, and die. The fruit is not affected. This fungus does not over-winter and is not carried by the seed. You should grow disease resistant varieties like:

Palmetto
Stono
Smoothie
Location: This fungus is found along the Gulf States and the Atlantic Seaboard.

MOSAIC (VIRUS)

The foliage mottles with a green and yellowish color, then curls. The fruit is bumpy and spotted. The total plant becomes stunted and reduction in yield follows. Mosaic also attacks musk-melons, squash, pepper, and tomatoes. This virus lives on the weeds and perennials around the garden such as milkweed, ground cherry, and catnip. It is spread by aphids to the cultivated plot.
Location: It is found throughout the United States and Canada. The best prevention is to do without the perennials above mentioned, and good control of aphids. Grow disease resistant varieties when possible. They are Ohio MR 200, Sensation Hybrid, Burpee Hybrid, Surecrop Hybrid, and Table Queen.

ROOT KNOT (NEMATODE)

The root sections of the plants are affected. At first tiny galls appear on the roots and as the roots grow the galls get bigger. Inside the galls are the white egg masses of the nematode. Severe damage can be done to a variety of crops. When the area is infested with nematodes the best possible thing you can do is rotate the whole garden to another site. If you only have one plot, treat the garden with a nematocide. This problem seems common only in the South.

SCAB (FUNGUS)

The fruit deveops a very dark brown spot which is covered with a gray fungus growth. A liquid runs out of the melon. The leaves and stems also develop spots. Severe damage can be expected in cool, damp weather. This fungus lives in the soil and is found in the Northeast and North Central States. Grow disease resistant Varieties such as Ashe, Highmoor, or Fletcher. Spraying for scab is a waste of time. Save your money.

PESTS OF EGGPLANT

COLORADO BEETLE (See Potato)

EGGPLANT LACEBUG

This gray or light brown insect gets its name from the lacelike wings of the adult. The nymph is yellowish and very small (one tenth of an inch).
Damage: Both adults and nymphs eat the foliage. They feed on the underside of the leaves. Leaves turn a yellowish-brown and the plants die.
Location: Only found in the South.

FLEA BEETLE (See Tomato)

HORNWORM (See Tomato)

DAMPING-OFF (See Celery)

FRUIT ROT (FUNGUS)

Brown lines appear at the soil line and brownish-gray spots on the stems and foliage. The fruit develops a large circular spot covered with blisters.
Location: It is found in all parts of the country. You can grow disease resistant varieties such as Florida Beauty or Florida Market.

WILT (FUNGUS)

The plants in the first stages look stunted, then slowly wilt and die.
Do not cultivate eggplant in soil that has been planted with potatoes in recent years. If you grow eggplant and potatoes, grow them at opposite ends of the plot. Location: Found in all parts of the country.

KALE (See Cabbage)

KOHLRABI (See Cabbage)

PESTS OF LETTUCE

APHIDS (See Muskmelon)

CABBAGE LOOPER (See Cabbage)

CUTWORM

Striped, with a Y on the front of the head. The color ranges from light green to blackish. About one and a half inches long.
Damage: Feeds mainly on grass crops but does some damage to lettuce and cabbage.
Location: East of the Rocky Mountains but not in cold regions. You can buy poison cutworm bait.

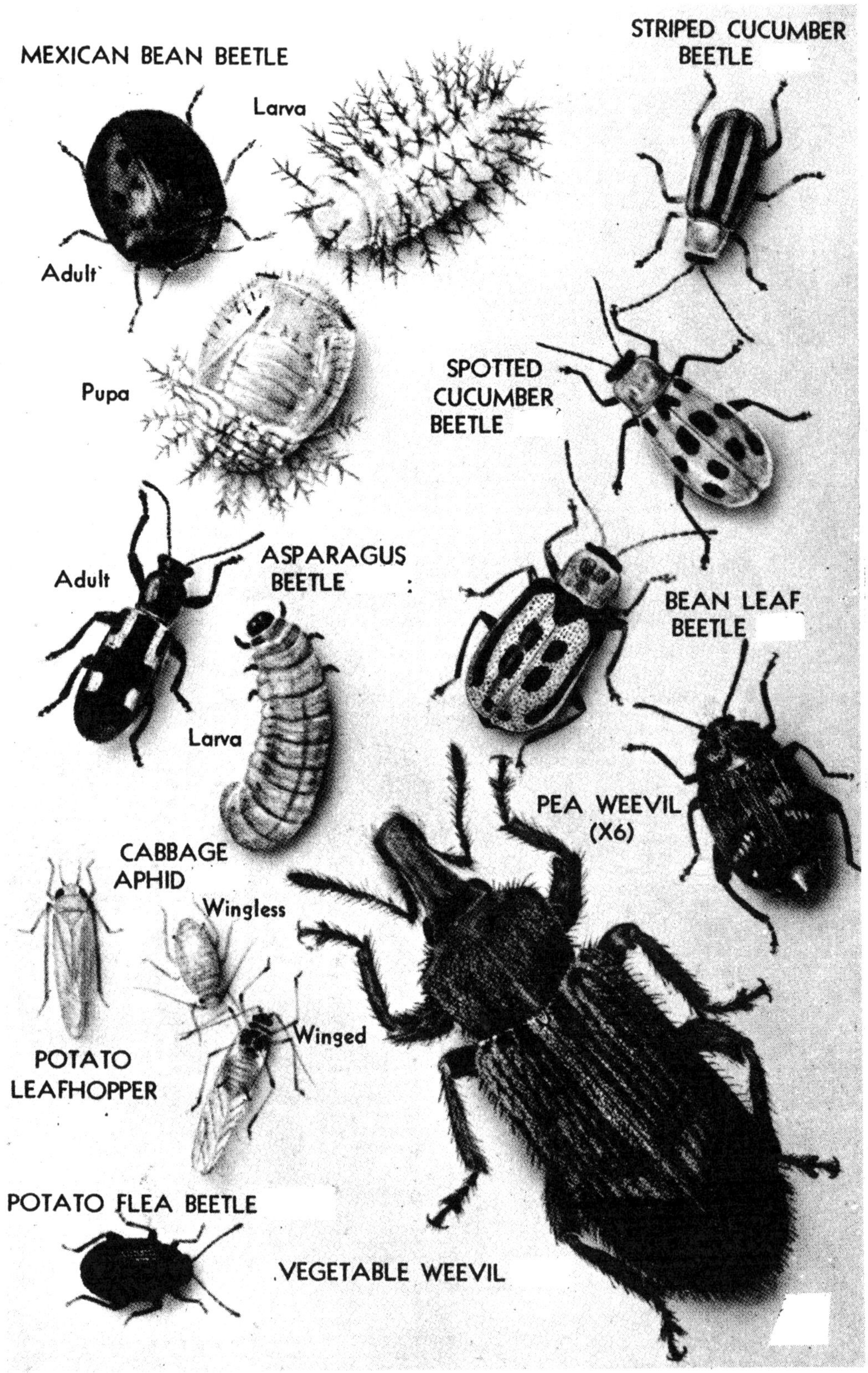
MEXICAN BEAN BEETLE
Larva
Adult
Pupa
STRIPED CUCUMBER BEETLE
SPOTTED CUCUMBER BEETLE
Adult
ASPARAGUS BEETLE
Larva
BEAN LEAF BEETLE
PEA WEEVIL (X6)
CABBAGE APHID
Wingless
Winged
POTATO LEAFHOPPER
POTATO FLEA BEETLE
VEGETABLE WEEVIL

SIX-SPOTTED LEAFHOPPER

Both the adult and young leafhopper are very light green and yellowish with small black dots on the wedge-shaped face. Very active and very small, it looks like a potato leafhopper.
Damage: Spreads the disease of aster yellows to lettuce, asters, and carrots. You must control the leafhopper or it will spread the disease from plant to plant. Try to plant lettuce in sheltered areas, well protected with screens.
Location: Found in all parts of the country.

WIREWORMS (See Potato)

DAMPING-OFF (FUNGUS) (See Cabbage)

PINK ROT (See Celery)

TIPBURN OF LETTUCE

The margins of the lettuce leaves turn brown and dry out. Head lettuce varieties have the most problems with this disease. You should grow disease resistant varieties like Cornell, Great Lakes, or Peenlake.

YELLOWS (VIRUS)

The inner leaves turn yellow and curl up. This is a common virus, also called aster yellows, and attacks both cultivated plants and wild plants. It is spread by the leafhopper.

MELONS (See Muskmelon)

PESTS OF MUSKMELONS

APHIDS

This is a large family of garden pests which damages almost all of the garden vegetables and many ornamentals- inside and outside. The adults and young are tiny and range in color from light green to black. These sucking insects cluster on the leaves and stems, usually on the underside.
Damage: The foliage becomes twisted and deformed. If the infestation gets out of control the plants may die. Many species of aphids carry plant diseases.
Location: They are found just about anywhere people cultivate plants.

PICKLEWORM

A brown head with a white-yellow body. The young worm has dark spots on its body. Measures up to three quarters of an inch long.
Damage: It feeds on the foliage and flowers, and will tunnel into the new buds and vines.
Location: Southeastern part of the United States but spreads north in warm weather.

SPIDER MITES

Spider mites are not insects but I list them here for they are a pest of most vegetable crops. You can hardly see them without the aid of a glass. There are several species and both the adult and the young damage a lot of plants. Their color ranges from red to greenish-red. They appear on the underside of leaves.
Damage: Little yellow dots and fine webs appear on the leaves. The plants and fruit may become stunted. Mites attack beans, blackberries, melons, and tomatoes.

Location: Found in all parts of the country.

Spider Mites

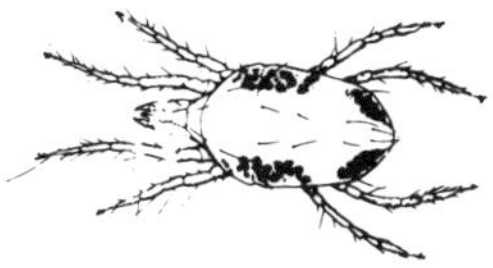

ANTHRACNOSE (See Cucumber)

BACTERIAL WILT (See Cucumber)

DAMPING-OFF (See Celery)

DOWNY MILDEW (FUNGUS)

Brown spots on the older leaves which dry, curl up, and finally die. The fungus does not affect the fruit. Grow disease resistant varieties: Delta Gold, Home Garden, and Gulfstream. Location: Found in all parts of the country.

LEAF-SPOT (FUNGUS)

The leaves develop small brown spots, but no spotting on the fruit. Location: Most parts of the country.

MOSAIC (VIRUS)

Mosaic diseases are caused by watermelon mosaic in the South and West, by tobacco ringspot virus in the Central States and East of the Rockies, and by cucumber virus and squash virus throughout the United States. The plants become a mottled green and yellow. The foliage curls, and the plant generally is smaller and deformed. Squash mosiac viruses are carried on the seed, tobacco ringspot virus lives in the soil, and watermelon mosaic virus is carried by the cucumber beetle. Keep the perennial weed crop down and spray for aphids and beetles. As soon as the disease appears, destroy the plant. Do not plant seeds from infested plants.

ROOT KNOT (See Muskmelon)

MUSTARD (See Turnip)

PESTS OF OKRA

CORN EARWORM

This worm can be either green, pink or brown. Is just over an inch long. You can handpick this worm. Damage: Eats holes in the pods. Location: Common in the South.

APHIDS (See Muskmelon)

JAPANESE BEETLE (See Sweet Corn)

STINK BUG (See Tomato)

ROOT KNOT (See Cucumber)

WILT

Leaves turn yellow and the plants do not develop. A serious problem in the South.

PESTS OF ONIONS

ONION MAGGOT

A white root maggot about one third of an inch long. Damage: Tunnels into the onion bulbs Location: Northern States.

ONION THRIPS

The adult does the damage. It is winged, very small, and is yellowish or light brown. Damage: Sucks the juice from the plant. Small white spots appear on the leaves and the tips turn brown and dry up. Location: Most of the country.

Onion Thrips

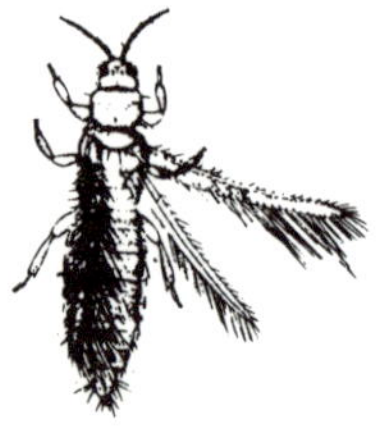

WIREWORMS (See Potato)

SMUT (FUNGUS)

Black blisters with small masses of fungus appear on the leaves. This fungus over-winters in the soil and will kill the young plants. If possible avoid all places where the fung has known to grow. Location: Found in the Northern Stat

ONION THRIPS

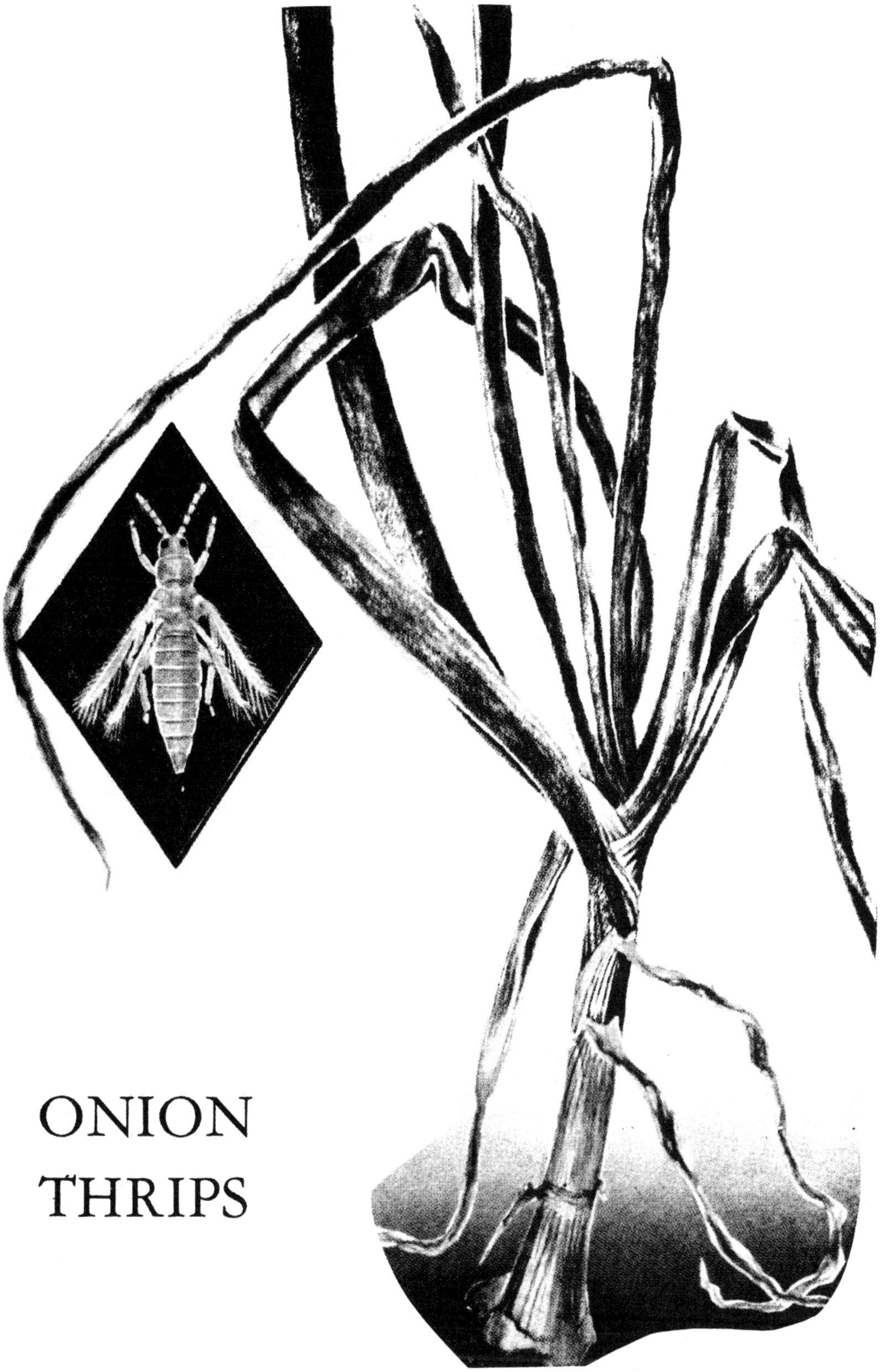

Onion plants showing severe thrips damage.
Insert, adult thrips.

PESTS OF BLACK-EYED PEAS

COWPEA CURCULIO

The adult is black with a hump-like back, and is, about one third of an inch long. The larva is legless and white with a yellow head.
Damage: Adults eat the pods and the peas. The larvae eat the green seed.
Location: South Atlantic States and Gulf States.

Cowpea Curculio

PESTS OF PEAS

APHIDS (See Muskmelon)

PEA WEEVIL

A light brown color with white and gray-black markings. About one fifth of an inch long. The larva is white with a brown head.
Damage: The adults feed on the flowers, and the eggs are laid on the pods. The larva tunnel into the pods.
Location: Throughout the United States and Canada.

ROOT MAGGOTS

Legless. About one third of an inch long when full grown. The color is yellowish-white. There are several species that include the corn maggot and the cabbage maggot.
Damage: Destroys seeds in the seed bed or garden plot. Tunnels into the root systems of many vegetables.
Location: Found just about everywhere.

ASCOCHYTA POD SPOT (FUNGUS)

Light colored spots on the pods with dark margins. The leaves develop a dark spot (very small). The stems develop spots near the soil line. This fungus is carried by the seed and lives on the remains of last year's plants. Ascochyta is found mainly in the Central States but has been seen in the Northeast.

BACTERIAL BLIGHT

Large spots on the pods that looked water soaked. The leaves then develop dark spots with a light colored, cream-like oozing effect. Bacteria are carried on seed and vines of last year's crop. Plant Western grown seeds for resistant plants.
Location: Found in most sections of the country except the semi-arid regions.

FUSARIUM WILT (FUNGUS)

The foliage and the interior of the stems turn a yellowish color. The plants wilt and, if severe, the fungus will kill the whole vine. It lives in the soil and enters the plant throu the root systems. You can grow diseas resistant varieties such as Alaska, Progress No. 9, World's Record, and Giant Stride.
Location: Found in all sections of the country.

ROOT ROT (FUNGUS)

The stems of the plant turn a brownish red to black and the whole plant takes on a sickly yellowish color. This blackness is usually beneath the soil line. Plants are the most susceptible at the time of flowering. A lot of moisture in the soil will encourage this fungus. Plant in a well drained soil.
Location: Most parts of the country.

Pea Weevil

A. Life stages (greatly enlarged): a. adult; b. pupa; c. larva; d. egg
B. Stem of pea vine with adult pea weevil in bloom, e.
C. Small pod with several eggs attached.
D. Larva in seed (greatly enlarged).

A. Aphid-infested pea plant showing damage and many aphids on the leaves and stems.
B. Undamaged pea plant
C. Aphids (greatly enlarged); a. winged adult; b. wingless adult; and c. half grown nymph.

PESTS OF PEPPERS

APHIDS (See Muskmelons)

CUTWORM (See Tomato)

EUROPEAN CORN BORER (See Sweet Corn)

FLEA BEETLE (See Tomato)

HORNWORM (See Tomato)

LEAF MINER (see Tomato)

ANTHRACNOSE (FUNGUS)

Dark brown spots or light black spots on the fruit. The centers have tiny black specks. This fungus is carried by the seed and lives in the soil. As in most fungus diseases, the best control is prevention. Plant treated seeds in a soil that has no history of fungus diseases. Anthracnose is found in most parts of the country.

BACTERIAL SPOT

New leaves develop a small greenish-yellow spot, with a darker yellow center. A dark margin forms on the older leaves, then they turn completely yellow and drop from the plant. The spots are about one quarter of an inch long. This bacteria, like most, is carried on the seed and over-winters in the soil.
Location: Found in all sections of the country and Canada except semi-arid regions.

BLOSSOM END ROT

The ends of the blossoms develop a very light colored, water soaked spot. As the flowers enlarge and the fruit develops, the rot will extend to the fruit and the fruit becomes dark and shriveled. A fungi may grow over the spot. To control this disease, avoid the overuse of fertilizer with a high nitrogen content and keep the soil at an even moisture level.
Location: Found in most parts of the country.

CERCOSPORA LEAF SPOT (FUNGUS)

Round water spots on the leaves and stems. The spots grow and sometimes become over one half an inch long with white centers and dark margins. The diseased foliage drop from the plant.
Location: It is most common in the Southern States.

DAMPING-OFF (See Beets)

MOSAIC (VIRUS)

The foliage turns a greenish-yellow and the leaves curl. The fruit develops a green ring spot. This is caused by the tomato and the cucumber mosiac. Try and grow these disease resistant varieties: Liberty Bell, Rutgers World Beater No. 13, or Keystone Resistant Giant. The above varieties are resistant to tomato mosiac but not to cucumber mosiac.

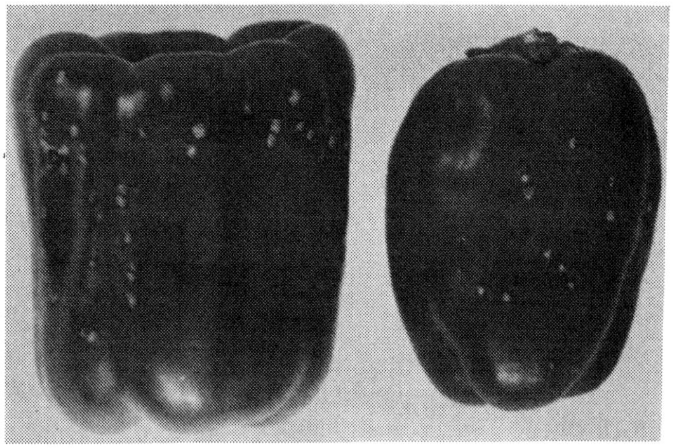

Bacterial Spot of Pepper

POTATO LEAFHOPPER

Potato leafhopper

a. adult leafhopper; b. nymphs; c. potato leaflets, showing upcurled brown tips and margins known as hopperburn, caused by the feeding of the leafhopper. (all enlarged)

CLARK'S NO. 1 POTATO.

PESTS OF POTATOES

APHIDS (See Muskmelons)

BLISTER BEETLE

This small beetle comes in many sizes, shapes and colors, which range from gray to striped. The body is slender and about one half to three quarters of an inch long.
Damage: Eats the foliage.
Location: Throughout the country.

COLORADO POTATO BEETLE

The adult is yellow with black stripes, and is about three-eighths of an inch long. The larva is brownish-red and about three-fifths of an inch long.
Damage: Both the larva and the adult eat the foliage of the potato, egg-plant, and tomato. They will completely defoliate small plants early in the season.
Location: East Coast

FLEA BEETLE (See Tomato)

GARDEN SYMPHLAN

This is not an insect but never-the-less a real pest. The adult has twelve pairs of legs but the young just a few. Any organic matter in the area probably will house some of these pests. They also like moist areas around potting sheds or home green-houses. They are white.
Damage: They eat the roots-small holes then larger ones indicate infestation. They damage beans, beets, carrots, celery, and spinach as well as potatoes.
Location: Most parts of the country but prefer moist, hot climates.

Garden Symphylan

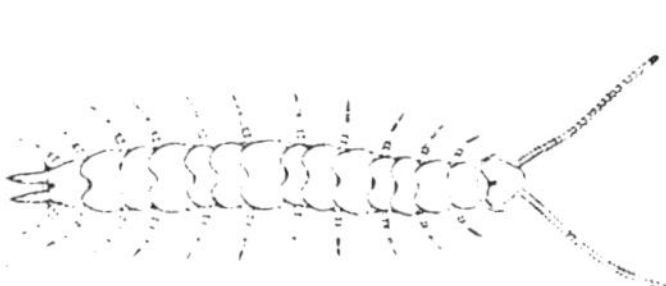

Grasshoppers

GRASSHOPPERS

I guess everyone has seen a grasshopper. If you haven't, you must live on the lower eastside of New York City, and you will probably find a few there. They are a problem for they are foliage eaters and do a lot of damage in certain parts of the country. See drawing for identification.

LEAFHOPPERS

The little green bug that flies when you brush up against a plant is probably a leafhopper. On close examination they are wedge shaped and very small. They crawl sideways like a crab.
Damage: Both the adult and the nymphs cause hopperburn of potato. The tips and the sides of the foliage turn a yellowish-brown and the leaves curl up.
Location: The Potato Leafhopper is found in the Eastern part of the country and the Western Leafhopper is found in the Southwest.

MOLE CRICKET (See Tomato)

POTATO TUBERWORM

A brown head with a white-pinkish body, about one half an inch long. If you hill the potatoes and keep out the weeds, it will help control this pest.
Damage: Tunnels in the tubers and stems. The plants wilt and sometimes die.

WHITE-FRINGED BEETLE

A snout and white marking along the side of the body clearly identifies this beetle. The adult is dark gray and the larva is yellowish-white. Both are about half an inch long.
Damage: Larva feed on the tubers and the root systems.
Location: Found in the Southeastern part of the country.

COLORADO POTATO BEETLE

a. adult beetle; b. eggs; c. larva (or slugs); d. pupa (or resting stage); (all enlarged).

WHITE GRUBS

All species have a hard brown head and range in color from white to light yellow. They are from about one half an inch long to one and a half inches. This is the larva of the May beetle. They live in the soil.

Damage: The grub eats the roots and the tubers and most underground parts of the plant.

Location: Most of the country.

White Grubs

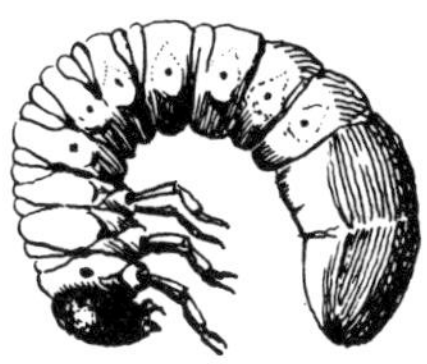

WIREWORMS

They are white to yellowish with darker heads and tails and up to one and a half inches long. They resemble a wire with joints.
Damage: Eats and tunnels into the tubers and roots of beans, carrots, lettuce, and turnips as well as potatoes.
Location: Most of the country.

COMMON SCAB (FUNGUS)

The scab appears on the tubers. Pitted and rough spots are a common symptom of this disease. This fungus is carried on the tubers and lives in the soil. It is found in most parts of the country. There are several controls for scab that involve spraying chemicals but the best is to grow disease resistant varieties such as Senaca, Early Gem, and Cayuga. You can help control this disease by planting clean tubers and by not using lime and fresh manure in the potato beds.

EARLY BLIGHT (FUNGUS)

The foliage develops a dark brown spot and looks like a target. Early blight fungus lives in the soil and is carried on the tubers.
Location: The distribution seems to be centered in the Eastern and South Central States.

LATE BLIGHT (FUNGUS)

The stems and leaves develop a dark area. The disease will kill plants early in the season. It increases activity during cool, moist weather.
Location: Northeast and Atlantic Coastal Regions. Grow disease resistant varieties such as Kennebec, Saco, Plymonth, or Pungo. If the tubers are infested they probably will rot in storage. Never dig up the tubers of diseased plants until the plant has died and dried.

LEAF ROLL (VIRUS)

The leaves roll up and the plant turns a pale yellow. Dark specks appear on the tubers. If you plant certified disease-free tubers, you probably will be safe. Saco and Katahdin varieties are resistant to tuber damage but the leaves and stems may become diseased. This virus is carried by tubers and aphids, so if you control them it will help keep this disease partly under control.

MOSAIC (VIRUS)

This virus is spread by aphids and the plant becomes stunted and the leaves curl. The mottled foliage is discolored and turns dark and light green. Grow plant resistant varieties like Saco, Sebago, Pungo, and Cherokee.

WILT AND DRY ROT (FUNGUS)

Stems and tubers develop brown rings and the leaves turn yellow and droop. Infested plants may die. If the tubers are infested they may rot in storage. The fungus lives in the soil and is carried on the tuber. Grow plant disease resistant varieties such as Menominee, or Ona. Never plant potatoes in soil that has had an infestation of fungus.

PUMPKIN (See Squash)

PESTS OF RADISHES

ROOT MAGGOT

About one quarter of an inch long and legless. The color is white and yellowish. There are many species that attack the roots of cultivated and wild plants. The maggot tunnels into the tubers.

PESTS OF RHUBARD

JAPANESE BEETLE (See Sweet Corn)

RHUBARD CURCULIO

A beetle with a yellow snout and about one and a half inches long.
Damage: Punches holes in the stems. These beetles breed on dock, so never plant rhubard near dock or remove dock from the garden plot.

Rhubarb Curculio

FOOT ROT OR CROWN ROT (FUNGUS)

Decayed stalks with brown spots at the bottom of the leaf stalks. The foliage wilts. This is a fast spreading disease so check it quickly. Remove all diseased parts as soon as possible.

PESTS OF SPINACH

APHID (See Muskmelon)

BEET WORM (See Beet)

LEAF MINER (See Tomato)

BLIGHT OR YELLOWS (VIRUS)

This is caused by the cucumber mosiac virus and is spread by aphids. The leaves curl and turn yellow. Some of the disease resistant varieties are Old Dominion, Virginia Savoy, and Varieties 56 and 612.

BLUE MOLD (FUNGUS)

A purple mold develops on the underside of the leaves and yellow spots on the tops. Most severe damage occurs during cool, moist weather. Found in the Southeast and Central States. Five disease resistant varieties are Dixie Savoy, Dixie Market and Hybrids 7, 424, and 425 (smooth leaf).

SEED DECAY (SEE PEAS)

PESTS OF SQUASH AND PUMPKIN

PICKLEWORM

About three quarters of an inch long with a brown head and a yellowish-white body. The young worms have dark spots on the body.
Damage: Feeds on most parts of the plant; buds, flowers, vines and fruit.
Location: Over-winters in the South and spreads north during hot weather.

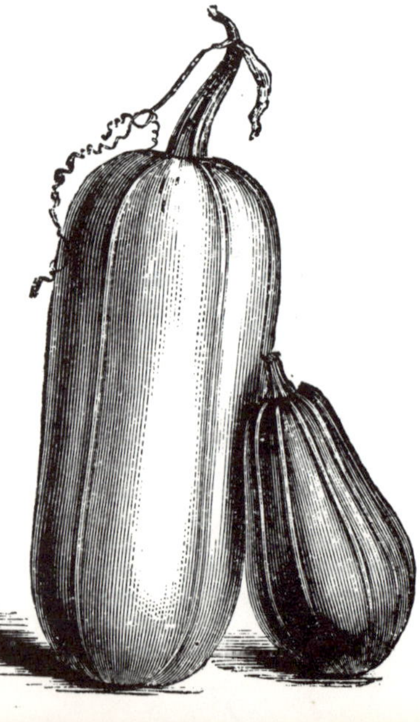

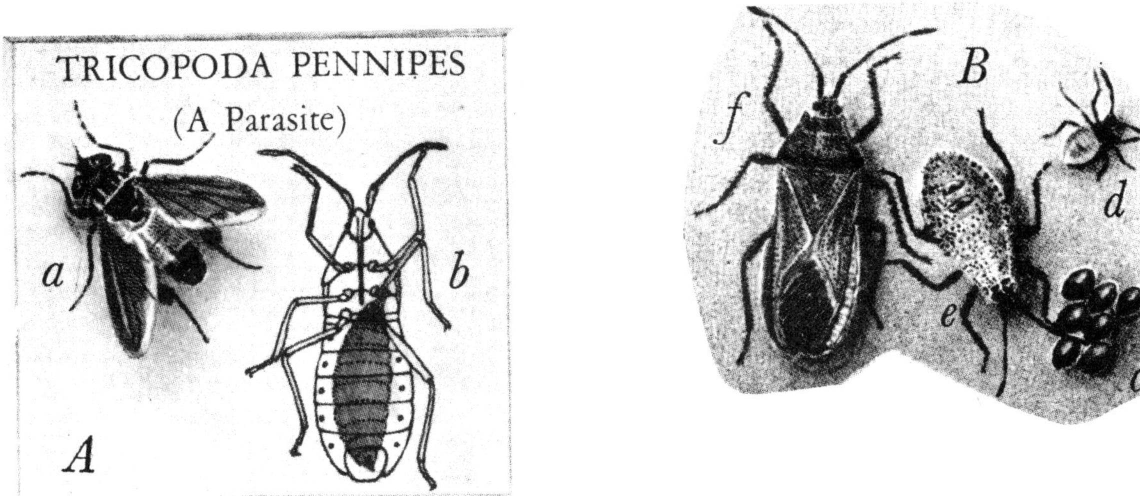

SQUASH BUG

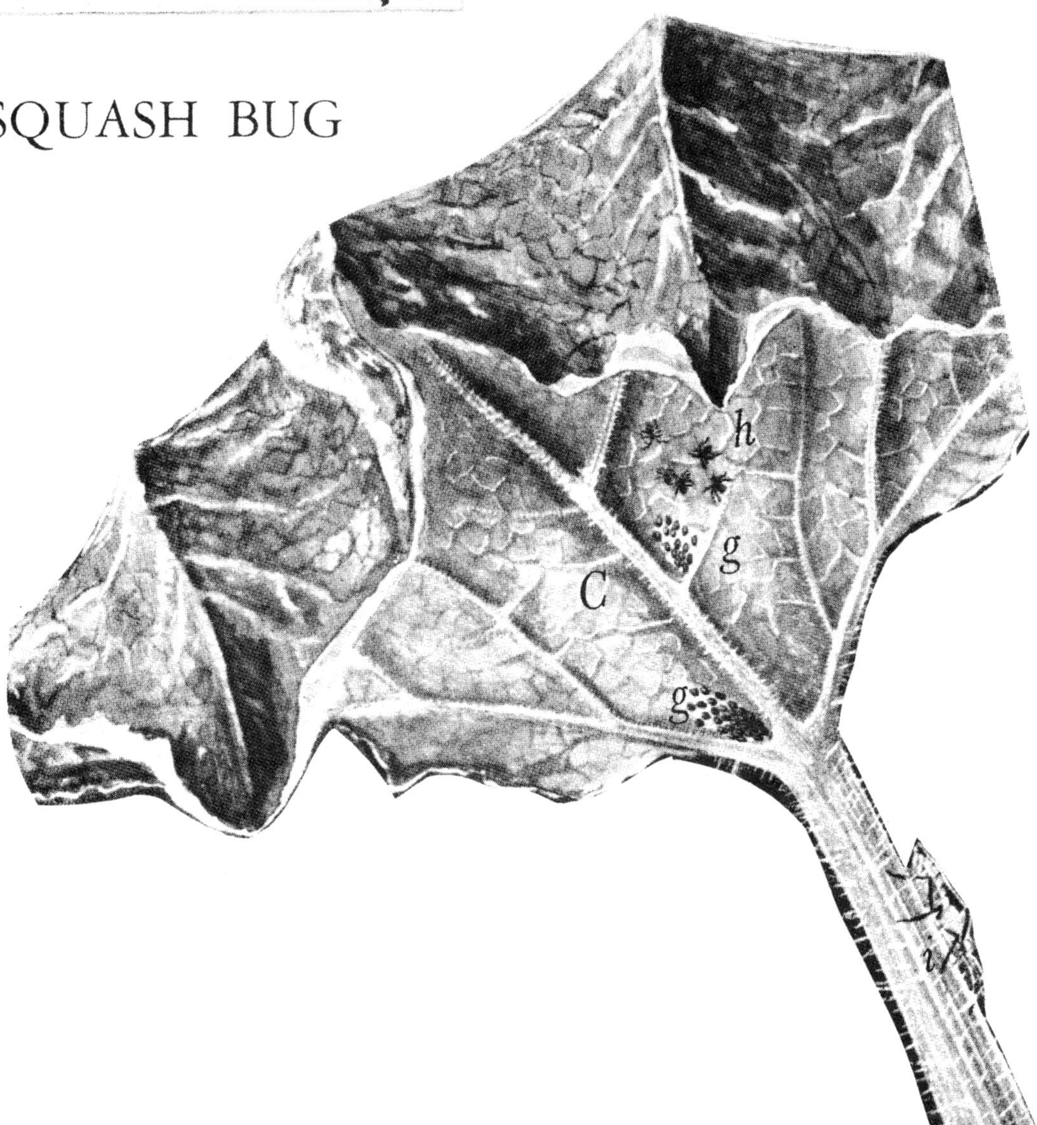

A, a, the parasite fly; b, diagram of a squash bug showing relative size and location of parasite larva in the host.
B. Some of the life stages of squash bug; c. eggs; d. newly hatched nymph; e. half grown nymph; f. adult;
C. Damaged plant; g. egg mass; b. nymphs a few hours old; i. adult. (all greatly enlarged).

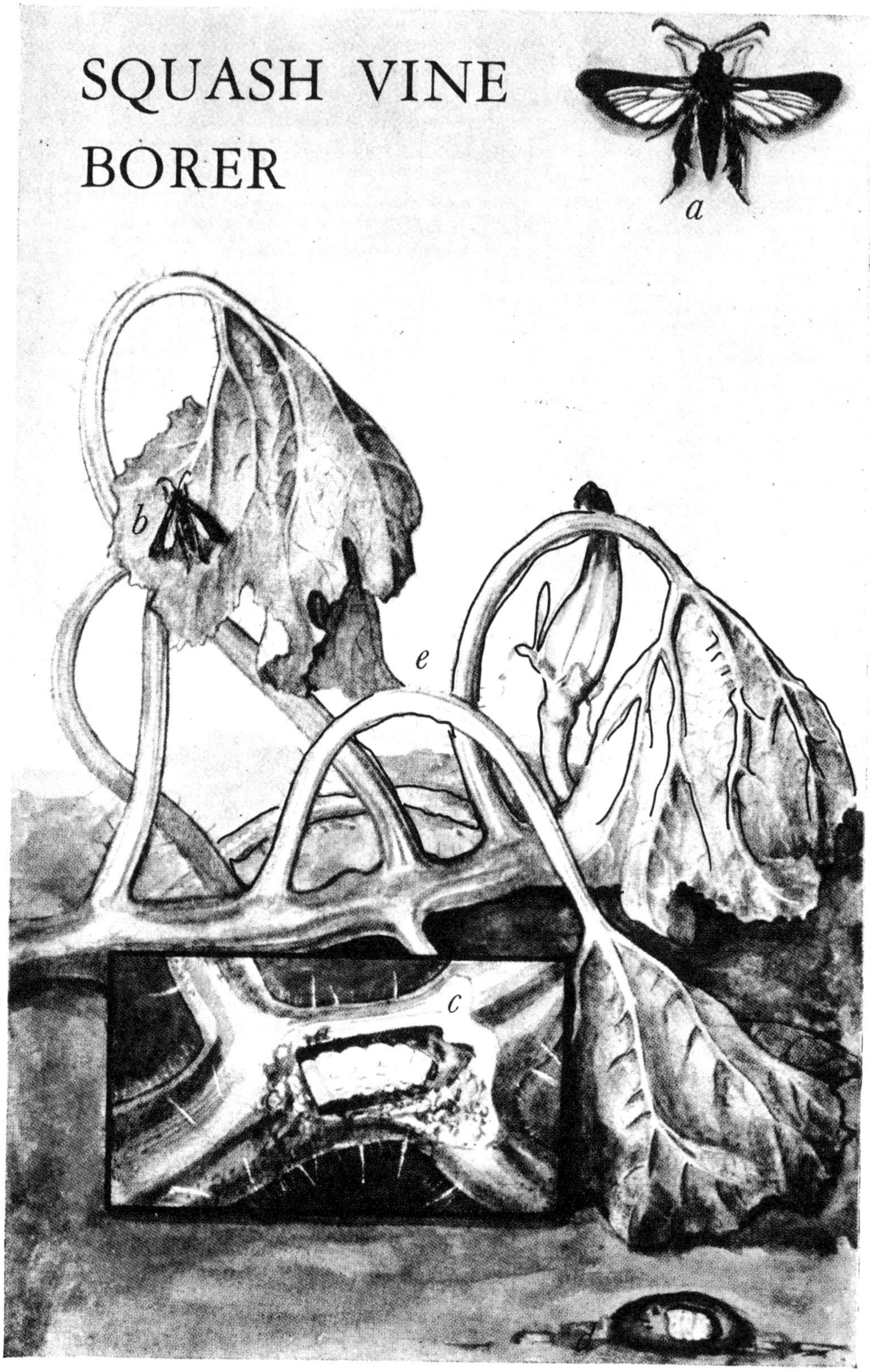

a. Moth (or adult) with wings spread; b. moth with wings partly folded; c. part of squash stem (enlarged) cut open to show borer (or larva) feeding: d. pupal cell in soil cut open to show pupa (or resting stage) inside; e. part of squash plant showing typical appearance of wilting caused by feeding of squash borer inside the stem.

SQUASH BUG

The adult is about three quarters of an inch long with a brown flat back. The eggs are red and nymphs range in color from a dark green with a red head to dark green-gray with a black head and legs.
Damage: A sucking insect that feeds in groups. You can handpick them easily from the plants. If the infestation continues the plant will die.
Location: Most of the country.

Squash Bug

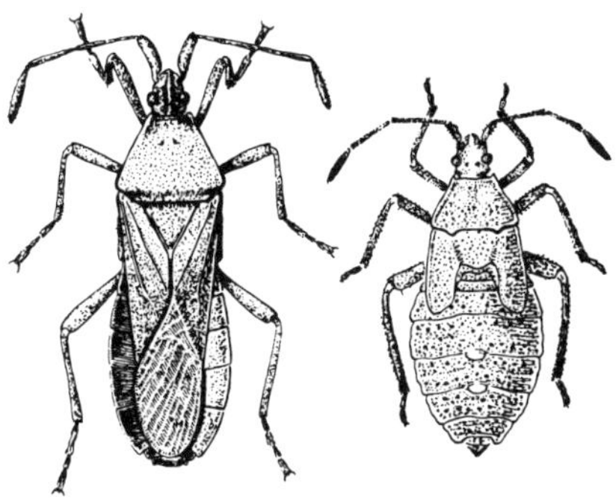

SQUASH VINE BORER

The larva is white and about one inch long. This insect is large enough to control by natural means. If you locate the spot on the vine where the injury is happening, split the vine, and kill the borer. After you remove the insect, mound up moist soil around the cut to prevent the vine from drying out.
Damage: Tunnels in the vines and eats the stem near the terminal shoots. The vines are held back.
Location: East of the Rockies.

Squash Vine Borer

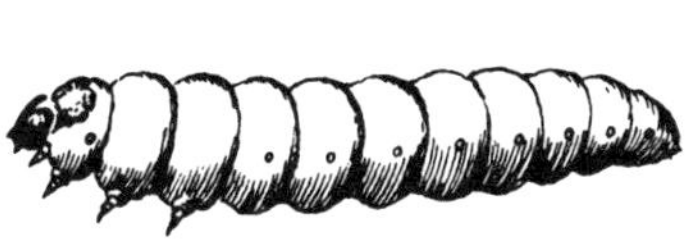

STRIPED CUCUMBER BEETLE

Yellow and black stripes down the back and just over one fifth of an inch long. The larva is slender and white, and about one third of an inch long.
Damage: The adult spreads the squash mosaic and bacterial wilt besides feeding on the foliage, stems and fruit. A real trouble maker. The larva tunnels into the root systems and sometimes feeds on the main stem just below the soil line. The plants wilt and die.
Location: East of the Rockies.

BACTERIAL WILT AND SCAB (See Cucumber)

MOSAIC (VIRUS)

The leaves and fruit develop yellow spots and the plants become stunted. Most commonly found on crookneck and straightneck varieties. With this mosaic, as with others, you must destroy and remove all infested parts and plants from the garden. Perennial weeds are a breeding ground, so remove weeds early in the season, before you transplant. Insect control also will help solve this problem as it is carried by aphids and the striped cucumber beetle.

PESTS OF SWEET CORN

CORN EARWORM

Brown, pink, or light green, with stripes down the side. Very large, about one and a half inches long.
Damage: During the first part of the season, this worm feeds on the shoots and causes budworm damage. Later on in the season, it tunnels into the silk and feeds on the fruit (kernels).
Location: Most of the country.

CORN SAP BEETLES

There are several species of this beetle in this country. The damage is done by the larva. You can identify the black adult before you see the white larva that eats the kernels. As in many cases the larvae are a bit larger than the beetle: In this case the larva is about one quarter of an inch and the adult three sixteenths of an inch.
Location: East Coast primarily, but has been seen in parts of Colorado.

EUROPEAN CORN BORER

Comes in three colors: whiteish, pink or brown with a dark head. The average length is just over one inch.
Damage: Eats the ears and stalks and damages the inside of the ear. This borer also feeds on pepper plants and will eat the fruit of the pepper.
Location: East Coast as far south as Georgia and west to Montana.

European Corn Borer

JAPANESE BEETLE

This pest was first discovered on the East Coast about 50 years ago and deserves special attention. A native of Japan, this beetle feeds on over two hundred plants native to North America. It is a quarter of an inch long with a metallic green back, copper-colored wings and has a grouping of white hairs along the sides of the back of the body. Both the male and the female have the same markings but the female is slightly larger. For the best part of the year (ten months), the beetle lives in the soil in the form of a white grub. This grub is the smallest of the white grubs that infest the grass regions of our country (our native white grub is a bit larger). As it lies in the soil, it rests in a position indicated by the photograph. Depending on the region, the beetles first appear on the foliage and begin eating in the late spring or early summer. A handy reference for your region is:

South...Mid May

Eastern...June 10th

Northern...July 1

Middle States...June 10th

Northern Central...June 30th

Japanese Beetle

When the sun is hot they become active and can be seen moving from plant to plant. This activity lasts for about five weeks, then the beetle gradually disappears. In the Northern States they hang around until frost, then begin the cycle again.

During the feeding periods, the female beetle tunnels into the soil and lays her eggs. After the egglaying process, she returns to feed with the adult male. If you are building a new lawn or importing topsoil for your garden, make sure the soil in not infested with the grub of the beetle, as this is a very common method of introducing the Japanese beetle into a new area.

Damage: The grubs feed on the tender shoots of all vegetables and grasses. The grub may go unnoticed until the plant begins to die or large areas of your lawn begin to turn brown.

EUROPEAN CORN BORER

a. Egg masses on underside of leaf; b. larvae in stalk and ear of corn; c. pupa in stalk; d. female moth; e. male moth; f. borings from burrow of larva.

The adult feeds on hundreds of different plants, from vegetables to ornamentals of all kinds. The adult beetle likes to feed in the bright sunlight and often can be seen clustering on plants during the very hot periods of the day. The foliage of plants resemble a lacy skeleton and, if serious infestations occurs, the plant may become completely defoliated. The beetles also mass on ripening fruit and will feed until nothing is left. They seldom bother unripe fruit. Serious damage is done to corn, for the beetle eats the silk as fast as it grows thus preventing the kernels from forming.

Controls: I have a section in this book for most of the controls of vegetables and fruit crops, but I have included here the controls for the Japanese beetle because of the extensive damage done and the vast amount of work and time spent by horticulturists on controls. There are many natural controls of the Japanese beetle, usually grouped under these headings:

INSECT PARASITES

Two very important parasites have been imported from Asia that prey upon this pest. The spring Tiphia, a wasp-like insect that lives on the grub and destroys it, and the fall Tiphia from Japan. Both of these insects are being dispersed by the government and are not available to commercial or home gardeners. It is being placed in wild places and in areas with heavy infestations of the beetle.

WEATHER

If the weather is very hot and dry in the summer months, the egg yield is increased and the adults are more active.

Japanese beetles congregating.

DISEASES

The milky spore disease lives on the grub and can kill generations of beetles when and if the spore takes hold. This spore causes the grub's clear blood to turn a milky color, and renders it helpless. This control is now available in garden centers. Under the heading of "Natural Controls," I have written at length on the Milky Spore Disease as a control for Japanese beetle grubs.

BACTERIAL WILT

The tassels turn white early, and the plants dwarf and wilt. A yellow slime oozes from cut stalks. This disease is carried by seed and other insects in the field. Found in the Eastern and South Central Regions. Grow disease resistant varieties such as Golden Cross Bantam, Spancross, Marcross, Stowell's Evergreen, and many of the late varieties. Try and buy seed that is disease resistant and save a lot of headaches.

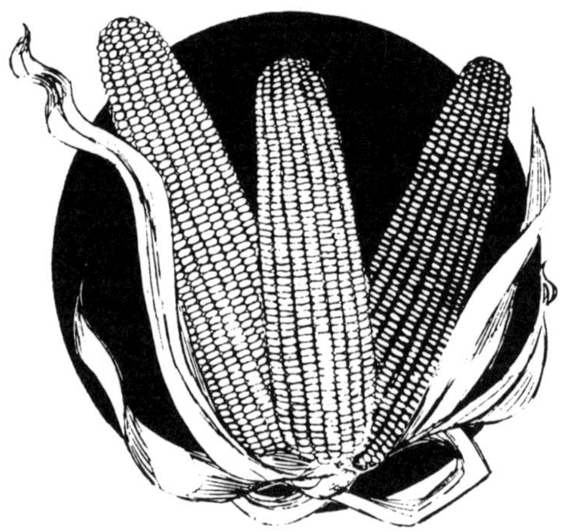

SEED DECAY AND SEEDLING BLIGHT (FUNGI)

The seeds decay in the soil and the young plants die. Found in most parts of the country. Buy treated seed.

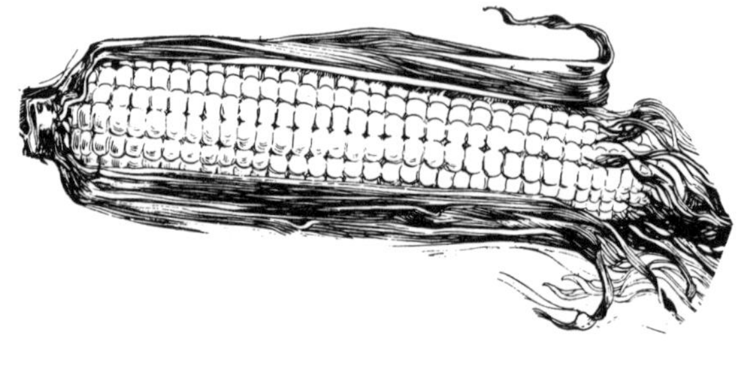

FALL ARMYWORM

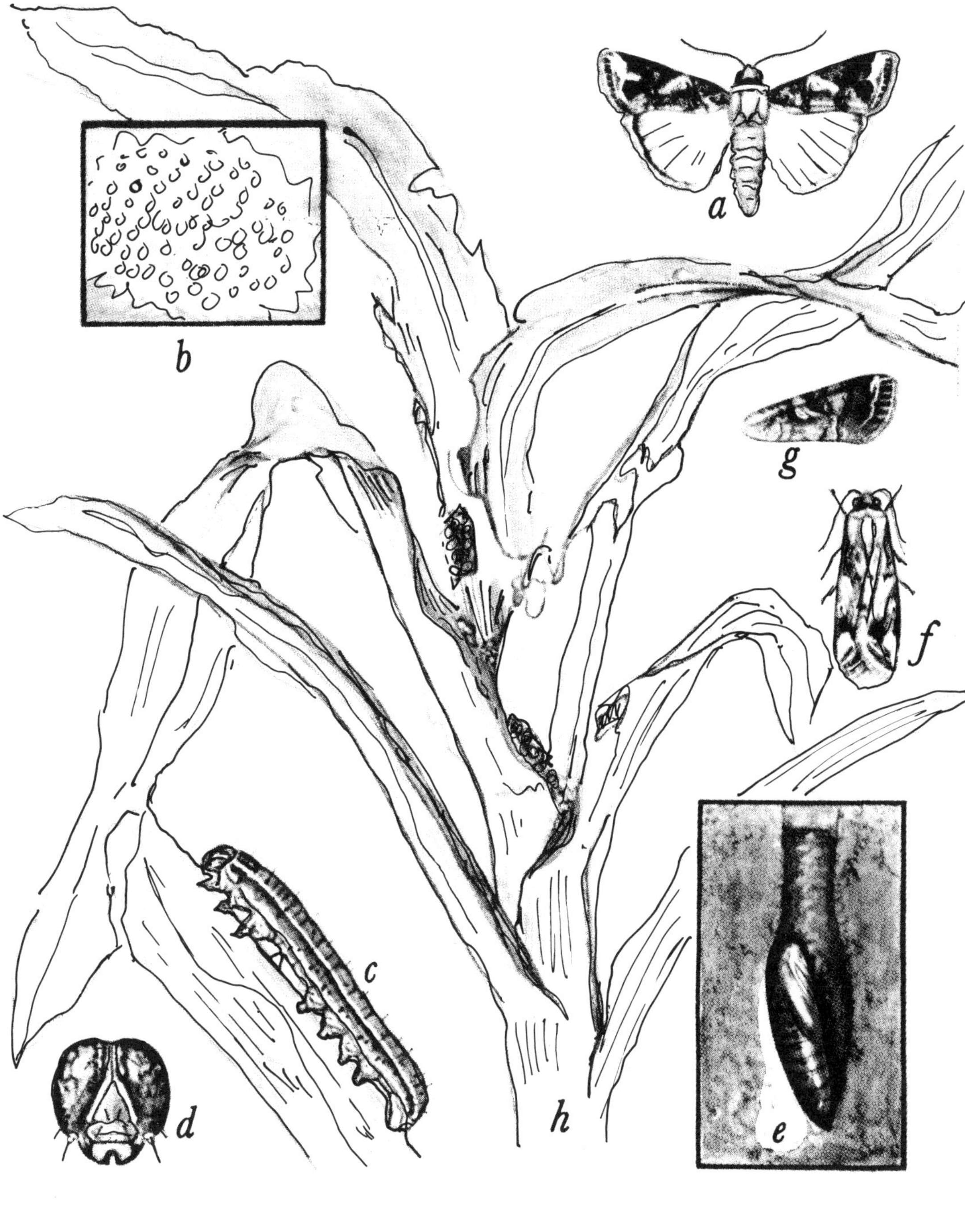

a. Male moth (or adult); b. eggs; c. larva; d. face of larva· e. pupa in cell; f. moth in resting position; g. wing of female moth; h. feeding injury to corn plant.

SMUT (FUNGUS)

White galls form on the ears and stalks of the plant, then burst and release masses of black spores. This fungus is common. Remove all diseased parts of the plant, and never put diseased parts in the compost bin.

PESTS OF SWEET POTATOES

SWEET POTATO WEEVIL

This beetle has a bluish-black head with a red snout and an antlike appearance. The body is shiny and about one quarter of an inch long. The larva is white with a dark blue head. It's about three quarters of an inch, and legless. This weevil does not hibernate so it must have a supply of food at all times. Destroy all plants in the garden after the growing season and plant weevilfree stock.
Damage: The adult weevil does little damage but the larva tunnels into the tubers and the vines.
Location: The South.

BLACK ROT (FUNGUS)

Round black spots that appear sunken on the tubers. Cankers on the underground stem sections. The fungus overwinters in the soil and is spread by the diseased tuber. Found in most parts of the country where sweet potatoe can be grown. Rotate sweet potatoes to a different part of the garden every two years.

STEM ROT OR WILT (FUNGUS)

If the plants begin to wilt and turn a yellowish color, cut the stems across --if they are black and discolored you probably have stem rot. This fungus affects the young plants and overwinters in the soil. The same treatment for black rot can be applied to stem rot.

Sweetpotato Weevil

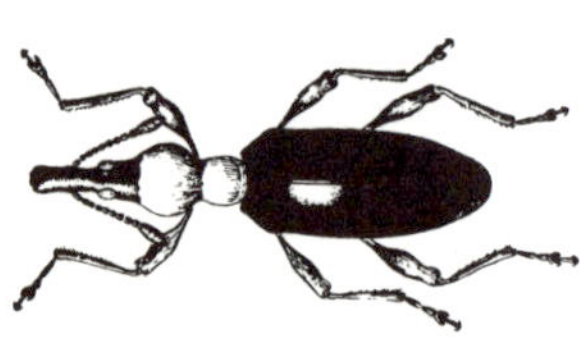

SWEETPOTATO WEEVIL

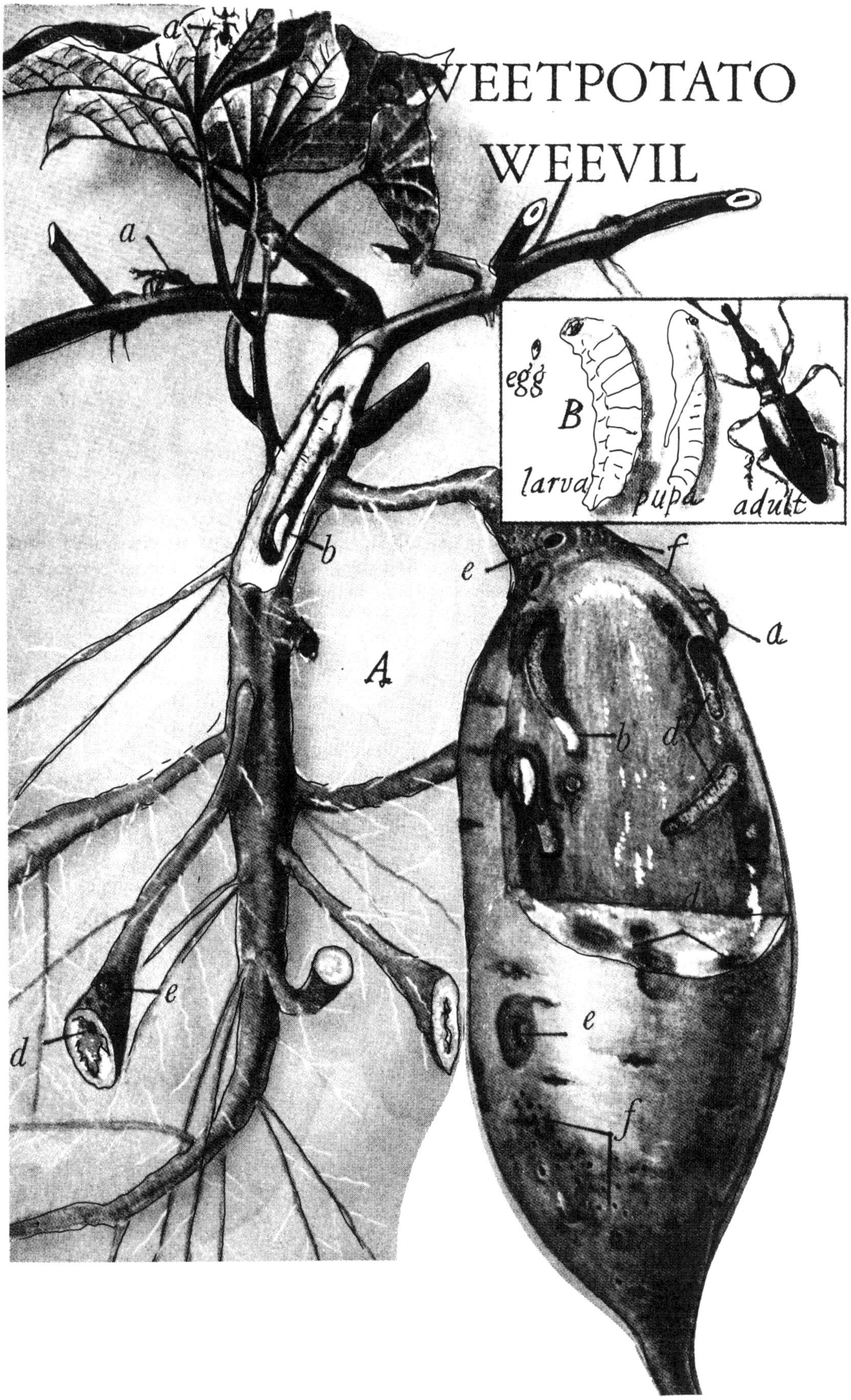

A. Weevil-infested sweetpotato and crown of plant; a. adults; b. larvae; c. pupa; d. larval injury; e. exit holes; f. feeding and egg punctures.
B. Developmental stages.

PESTS OF TOMATOES

APHID (See Muskmelon)

CUTWORM

If you have never seen a cutworm on your tomato plants, consider yourself lucky. This is a large worm that eats and eats. It comes in many species but usually is gray and may be spotted.
Damage: Eats the foliage and cuts the plants in two spots-above the soil line and below the soil line. Real damage can be done to young tomato plants as well as pepper and bean plants.
Location: Most of the country.
One natural control of cutworms, which works, is mechanical. Place a cardboard collar around the stem of the plant. Push the stiff board down in the soil about one inch so that three inches are above the soil. The collar should be about two inches from the stem to allow it to grow. The collar keeps the worms from climbing onto the plant and prevents them from cutting the stems.

FLEA BEETLE

A very small jumping beetle. Many species. It is striped and usually black or brown.
Damage: Eats the foliage of many vegetables and, if not checked, will do a lot of damage. The leaves look as if they have been shot with buckshot.
Location: Throughout the country.

HORNWORMS

This is the big green worm (sometimes four inches long) with a horn on the rear end, that eats the leaves and the fruit of the tomato, pepper, and eggplant. Handpick these pests. They seem to follow the tomato crop.

LEAF MINERS

The adults are tiny flying yellow and black bugs. The larva is yellow and very small. If not controlled early in the season, several generations may develop.
Damage: The larva tunnel through the leaves of tomato, pepper, and spinach. The tunnels have white markings.
Locations: Damage is most severe in the hot regions but it is found in most parts of the country.

Hornworms

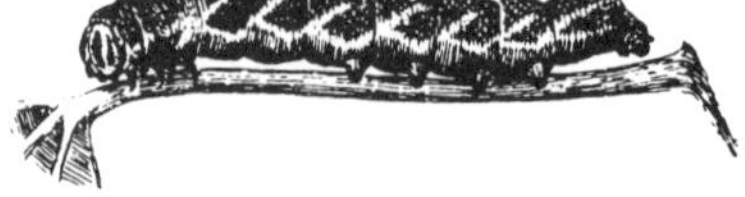

HORNWORMS ON TOMATO

a. Tomato hornworm moth (or adult) with wings stread; b. egg; c. larva, dark form; d. pupa (or resting stage); e. tobacco hornworm larva, light form (a. enlarged);

TOMATO FRUITWORM

Tomato fruitworm

a. Female moth with wings spread (adult); b. male moth with wings in natural position; c. eggs; d. larva; e. pupa in its cell in the soil; f. larva feeding on tomato fruit, showing typical injury. (all enlarged).

STINK BUGS

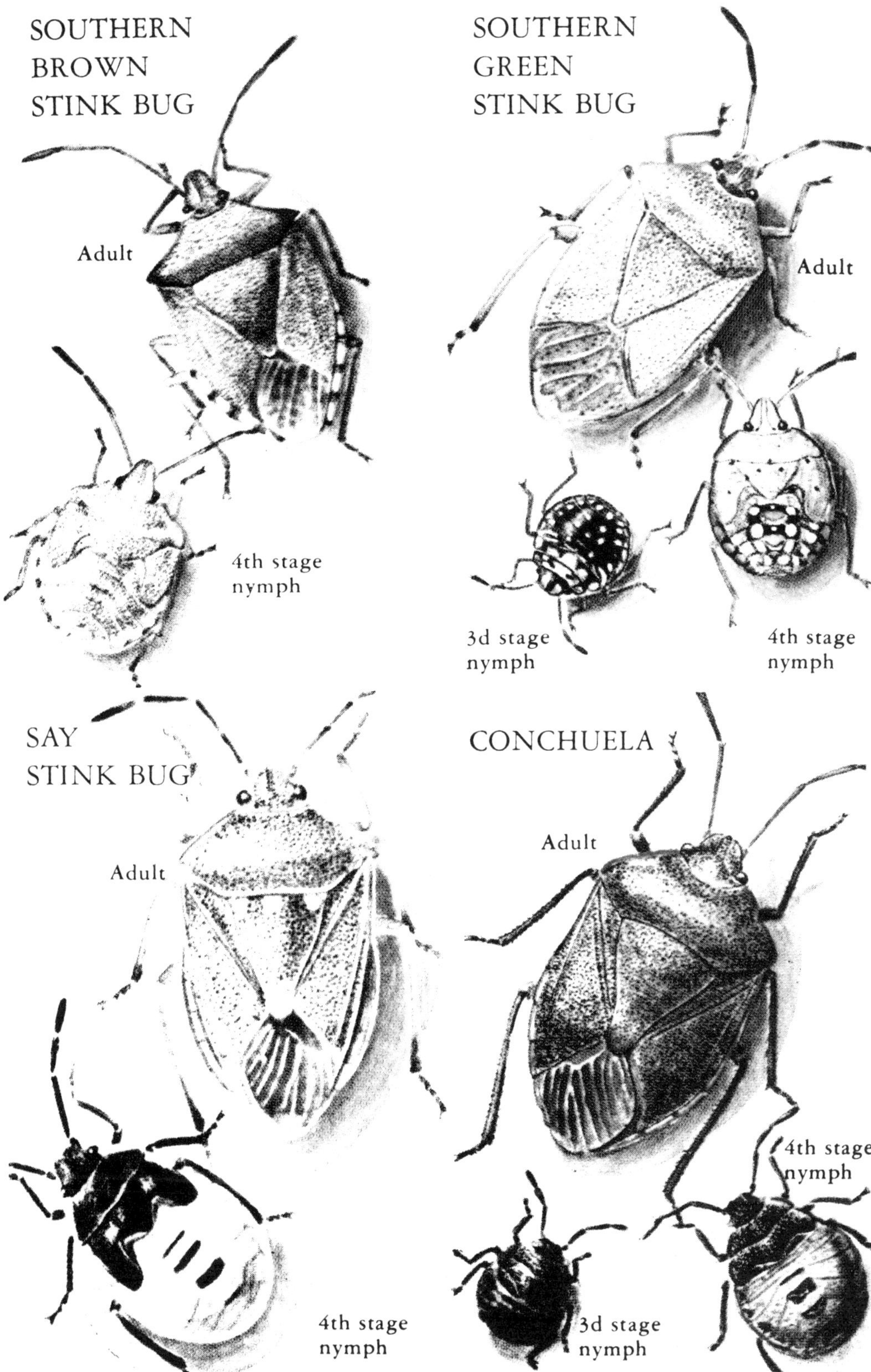
SOUTHERN BROWN STINK BUG
Adult
4th stage nymph
SOUTHERN GREEN STINK BUG
Adult
3d stage nymph
4th stage nymph
SAY STINK BUG
Adult
4th stage nymph
CONCHUELA
Adult
4th stage nymph
3d stage nymph

MOLE CRICKETS

The adult and nymphs do a lot of damage in the seed beds. The adult is about one and a half inches long with stout legs and beady eyes. Light brown in color with shovel-like feet.
Location: The South.

Mole Crickets

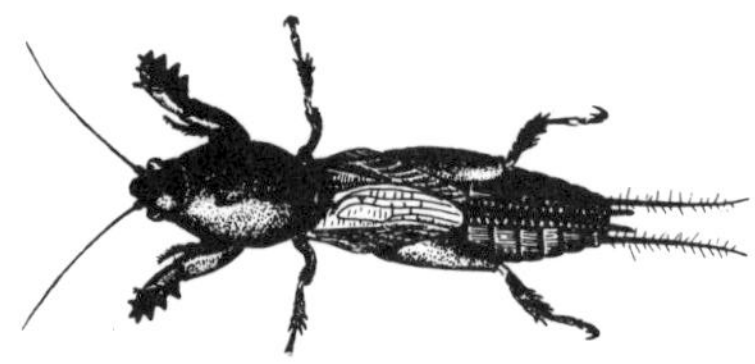

SPIDER MITE (See Muskmelon)

STALK BORER

A slender worm, creamy in color with a dark saddle around the body. Some species have purple stripes down the body. This insect breeds on weed plants, so destroy all weeds in and around the garden plot. If you can locate the borer in the stem before a lot of damage is done, you can puncture the insect with a hatpin.
Damage: Tunnels in the stem and kills the plant. The opening where the borer enters the stem is one quarter of an inch.
Location: East.

Stalk Borer

STINK BUG

A sucking insect that comes in several colors-green, brown, or black. It is shield-shaped and some have markings on the back. They are one half an inch to one third of an inch long. A foul odor is discharged from this pest, thus "stink bug". The nymphs are the same shape and color, only smaller.
Damage: Buds and young fruit are malformed as the bug, in both the adult and nymph stages, suck the sap from the plant. Besides attacking tomatoes, they suck on bean and okra plants, causing wart-like growths to appear on the pods.
Location: Mostly in the South, but has been seen in parts of the North.

TOMATO FRUITWORM

About one and three quarters of an inch long. Comes in various colors-green, brown, or pink, with light stripes along the side. Damage: Eats holes in the tender buds and the fruit of the plant.
Location: Heavy concentration in California and the South but found in most parts of the country.

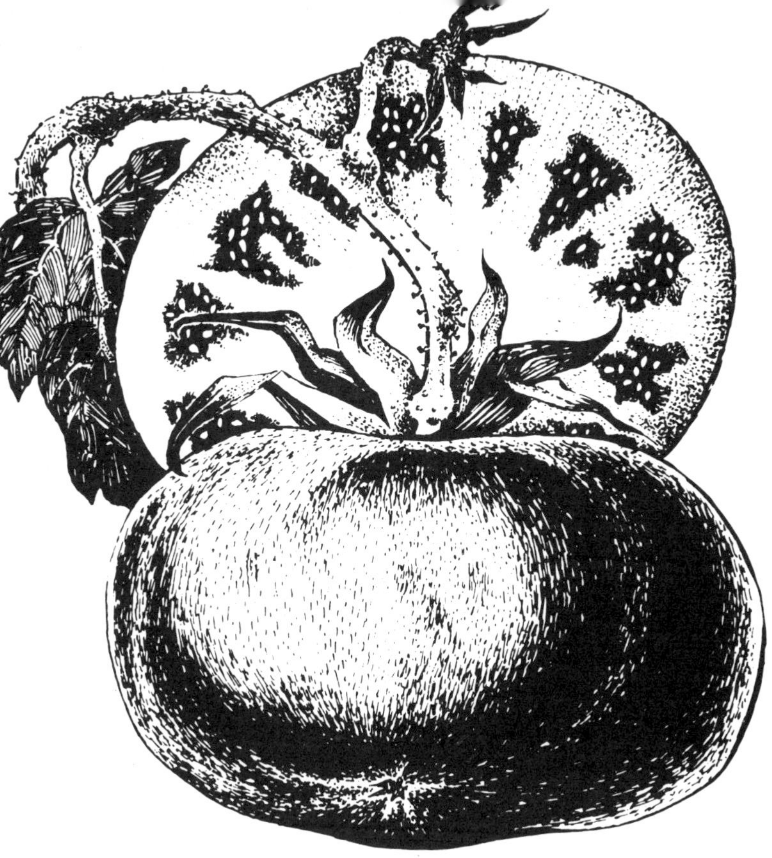

TOMATO RUSSET MITE

This pest cannot be seen with the naked eye, but if you use a twenty-power lens you will be able to see the pear shaped masses on the stems. The stems take on a smoky color. The underside of the foliage is also affected. Location: Found in California but has been seen in remote cases in other parts of the country.

BLOSSOM END ROT

The dark leathery spots which appear at the end of the fruit are caused by a calcium deficiency. Most damage occurs during hot weather when the plant is trying to make fast growth. Use lots of lime and keep the soil moisture even throughout the season. Found in most sections of the country. Avoid heavy use of nitrogen fertilizers if this condition prevails.

DAMPING-OFF (See Beet and Chard)

EARLY BLIGHT (FUNGUS)

Dark brown, irregular spots which often enlarge into target-like markings on the fruit and the ends of the stems. The stems develop cankers that may girdle the plant at soil level. The fungus lives in the soil and does the most damage during hot, humid weather. Location: Found almost everywhere except in arid regions.

FUSARIUM WILT (FUNGUS)

The fungus starts at the bottom of the plant and works its way up to the tips. The foliage gradually turns yellow and wilts. The stems and woody parts turn brown. The plant will die if the disease is not checked and the bad parts removed. The fungus lives in the soil and enters the plant through the root systems. You can grow disease resistant varieties such as Manapal, KC 146, Marion, Homestead, Porte, H 1350, Pinkshipper, Kokomo, and others (See Vegetable Chart).
Location: Found mostly in the South.

LATE BLIGHT (FUNGUS)

Water soaked spots, usually dark on the foliage. The fruit develops a dark spot and a white fungus forms on the underside of the leaves. As the disease progresses the spots on the fruit (see photograph) turn brown and black. The leaves wilt and turn brown. This usually appears during hot weather.
Location: East of the Mississippi.

LEAF SPOT (FUNGUS)

Leaves develop a small spot that has light centers and dark margins. Small specks appear in the center of the circle. This fungus spreads during hot weather and lives in the soil and on weeds.
Location: Found in the Northeast and North Central States.

MOSAIC (VIRUS)

The leaves curl and the foliage turns a mottled green and yellow. This mosiac is caused by tobacco mosiac and sometimes is present on manufactured tobacco products. It can be transmitted by hand while smoking in the garden or by dropping cigarettes in the garden. It is also carried by aphids. This virus does not transmit by seed and does not live in the soil. You can prevent this disease by not handling the plants (especially the young) if you smoke or are smoking. Wash you hands before transplanting the seedlings or just working around tomato plants.
Location: Found in all parts of the country.

ROOT KNOT (NEMATODES) (See Cucumber)

VERTICILLIUM WILT (FUNGUS)

This disease resembles fasarium wilt. The big difference is verticillium symptoms appear on the whole plant at the same time. The plants will wilt and the foliage turns yellow. The fruit may sunburn if exposed to the sun after the disease has attacked the plant. Grow disease resistant varieties such as Enterpriser, Campbell's 17, H 1350, Porte, and Person Types VF6 and VF36. These varieties are also resistant to verticillium wilt only. Pesticides are not effective against verticillium wilt.
Location: Found on the West Coast, North Central, Northeastern, and Southern Florida.

PESTS OF TURNIP AND MUSTARD

CABBAGE LOOPER (See cabbage)

FLEA BEETLE (See Tomato)

HARLEQUIN BUG

A brightly colored bug with a red and yellow shield. The body is black and about one half an inch long.
Damage: Plants wilt and the foliage turns brown.
Location: Southern States and California

ROOT MAGGOT (See Peas)

TURNIP APHID (PLANT LOUSE)

A tiny green aphid, without the usual wax coat. Feeds on the underside of the leaves.
Damage: A sucking insect that causes the leaves to curl and turn brown.
Location: Most of the country except the Northwest.

VEGETABLE WEEVIL

Grayish-brown adult with two oval pale marks on the back. Just over a half inch long. The larva of the species has a greenish body with a light yellow to tan head.
Damage: Adults and larvae both feed on the foliage and roots of the turnip, cabbage, and related plants.

WIREWORMS (See Potato)

PESTS OF WATERMELONS

Aphids, spider mites, and the striped cucumber beetle all have been covered under other vegetables. These are the main pests of the watermelon.

ANTHRACNOSE

Dark spots on the leaves and fruit. The fruit also develops water-soaked spots which have a dark center. A pinkish fungus growth appears in the hot weather. The spots on the fruit are small and raised at first but become larger and sunken in appearance as the disease progresses. This fungus also attacks cucumber and muskmelon. It is carried on the seed and lives in the soil. Grow disease resistant varieties: Fairfax, Charleston, Gray Congo and Blackstone.

WILT

The seedlings are stunted and the vines wilt. The disease starts at the tips and spreads to the whole plant. Carried by seed and lives in the soil. Grow these varieties: Improved Stone, Mountain No5, Fairfax, Kleckley No6, Charleston Gray, Hawksbury, Missouri Queen, Klondike R-7, and Baby Klondike.

KNOT ROOT (NEMATODE)

VEGETABLE	SEEDING DATE	APPROXIMATE TIME OF HARVEST	SUGGESTED VARIETIES
Snap beans	June 30-July 15	August 20-Sept. 15	Provider Spartan arrow
Beets	July 1-15	September 1-October 1	Ruby queen Detroit dark red
Broccoli	June 20-30	Sept. 20-October 20	Green comet Spartan early
Brussels sprouts	June 20-30	October 15-November 1	Jade cross Long Island improved
Cabbage	June 20-30	October 1-October 30	Stonehead, Ruby ball (red), C-C cross
Cailiflower	June 20-30	October 1-October 30	Self-blanche Snowball imperial
Chinese cabbage	June 20-July 30	October 1-October 30	Early hybrid G Springtime
Kale	June 20-July 30	Aug. 15-Sept. 30	Dwarf blue curled Vates
Kohlrabi	June 20-30	October 1-15	Early white vienna
Lettuce (butterhead)	July 15-30	Sept. 15-October 15	Butter-crunch
Lettuce (leaf)	July 1-Aug. 15	Aug. 15-October 10	Salad bowl Grand Rapids
Peas	July 15-July 30	Sept. 15-Oct. 10	Freezonian, Greater progress Dwarf gray sugar (edible pod)
Radishes	Aug. 1-Sept. 1	Sept. 1-October 1	Cherry Belle Champion
Spinach	July 20-30	Sept. 1-20	America Viking
Turnips	July 1-Aug. 1	Aug. 15-October 1	Tokyo cross Just right

DISEASES

Black rot of cabbage

Early blight of tomato

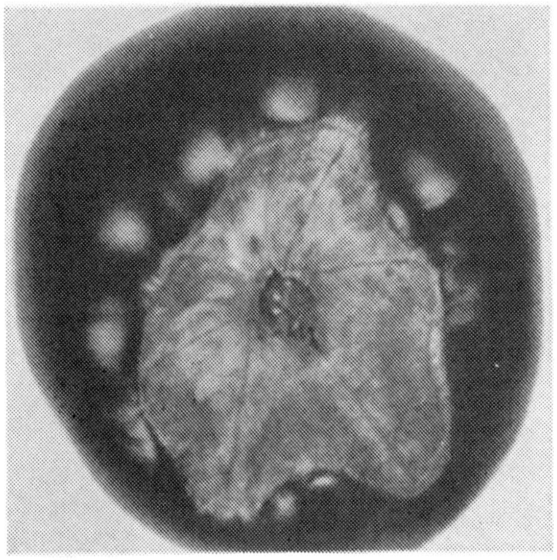

Blossom-end rot of tomato

Bean rust

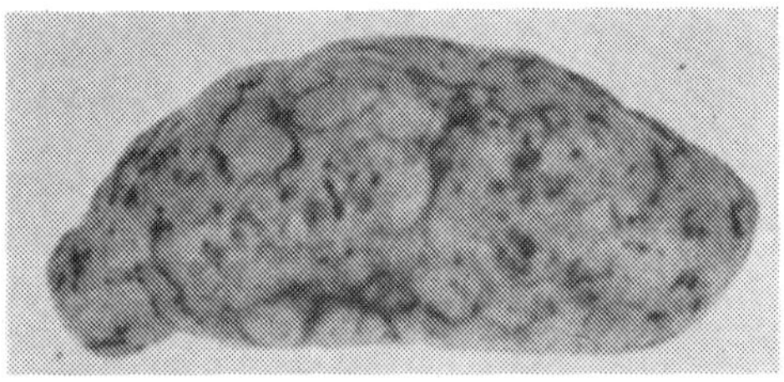

Mosiac virus disease of squash

Potato scab

Cucumber mosiac

Fusarium wilt of tomato

omato leaves showing
ymptoms of early blight.
ote target-like markings.

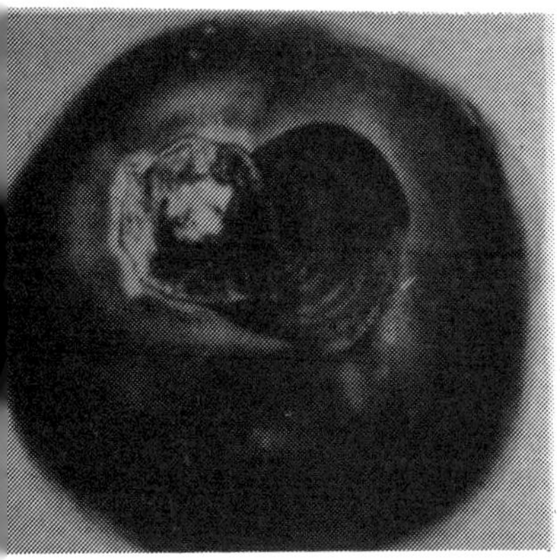

omato affected with early
light.

Tomato leaves showing dark
blotches characteristic of
late blight.

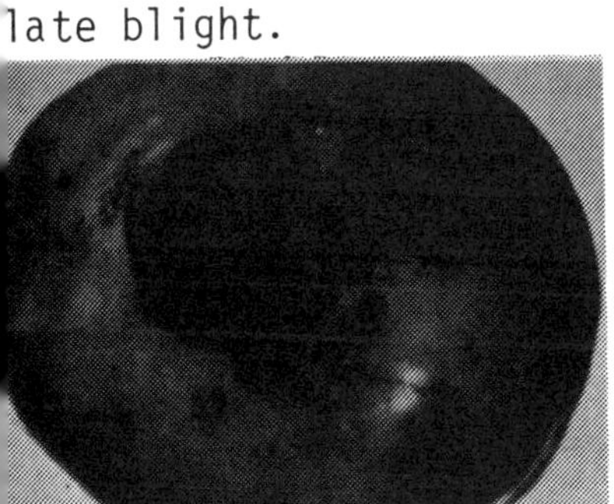

Tomato showing symptoms of soil rot. Note broken tissue and targetlike markings.

Tomato showing symptoms of catface. Note cellular malformation of tomato.

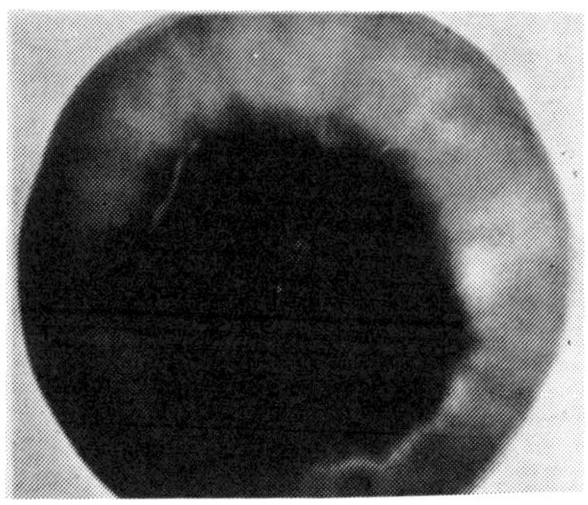

Tomato showing symptoms of blossom-end rot. Spot is dark, sunken, and leathery.

Tomato leaf showing downward curling that is characteristic of 2,4-D injury.

Tomato leaf showing mottling (yellow) caused by tobacco mosaic.

Tomato plant showing bacterial wilt. Note rapid wilting of entire plant.

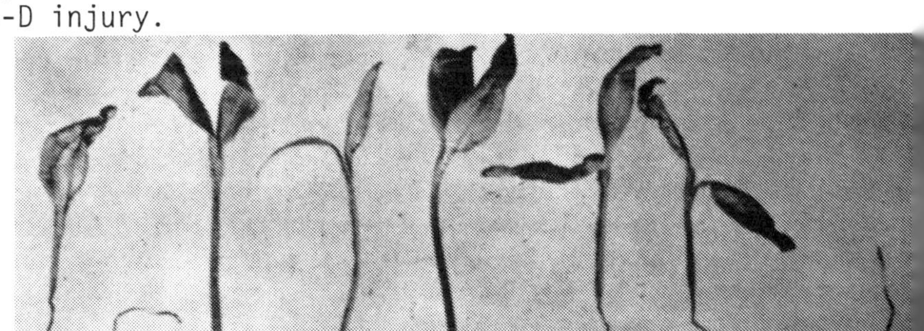

Tomato seedlings affected by damping-off.

VEGETABLE CHART OF VARIETIES SUITED FOR THE HOME GARDEN

Variety	*Remarks*
Asparagus	
Mary Washington	Widely adapted, tolerant to rust
Beans (bush)	
Topcrop	Widely adapted, resistant to bean mosaic
Contender	Resistant to bean mosaic and powdery mildew
Bush Blue Lake 274	Heavy yield, resistant to bean mosaic
Resistant Cherokee Wax	Yellow pods, resistant to bean mosaic
Resistant Kinghorn Wax	Yellow pods, resistant to bean mosaic
Beans (pole)	
Kentucky Wonder	Widely adapted, excellent for freezing
Dade	Adapted to South, resistant to rust, common and Southern bean mosaic
Romano	Widely adapted, good for freezing
Beans, lima	
Fordhook 242	Widely adapted
Henderson's Bush	"Baby" or small seeded
Jackson Wonder	Widely adapted
Beets	
Detroit Dark Red	Widely adapted
Ruby Queen	Widely adapted, good for canning
Broccoli	
Waltham 29	Widely adapted, good for freezing
Green Comet	Widely adapted
Brussels sprouts	
Jade Cross	Widely adapted
Long Island Improved	Heavy yield over extended period
Cabbage	
Golden Acre	Resistant to cabbage yellows
Early Jersey Wakefield	Slow bolting, resistant to cabbage yellows
Copenhagen Market	Widely adapted, early maturity
Red Acre	Red color
Chinese cabbage	
Michihli	Widely adapted
Carrots	
Danvers 126	Widely adapted, heavy yield of long tapered roots
Royal Chantenay	Widely adapted, broad-shouldered roots with little taper
Imperator Long Type	Widely adapted, long slender roots
Cauliflower	
Snowball Y	Widely adapted, produces over a long period
Early Snowball A	Early maturing
Celery	
Utah 52–70	Widely adapted, tolerant to celery mosaic, resistant to boron deficiency
Collards	
Vates	Slow bolting
Georgia	Widely adapted
Corn, sweet	
Iochief	Resistant to Stewart's wilt
Seneca Chief	Resistant to Stewart's wilt
Butter and Sugar	Mixture of white and yellow kernels, excellent quality
Golden Security	Resistant to Stewart's wilt
Cucumber (slicing)	
Ashley	Resistant to downy mildew
Saticoy Hybrid	Tolerant to downy mildew and mosaic
Marketer	Resistant to downy mildew
Cucumber (pickling)	
Pioneer	Tolerant to scab, mosaic, downy and powdery mildew, anthracnose, and angular leaf spot
Wisconsin SMR–18	Resistant to scab and mosaic
Eggplant	
Jersey King	Elongated fruit, widely adapted
Black Beauty	Oval globe fruit, widely adapted
Mission Bell	Early high yield
Kale	
Vates	Slow bolting
Siberian (sprouts)	Very vigorous
Kohlrabi	
Early White Vienna	Widely adapted

Tomato leaf showing symptoms of gray leaf spot. Spots are small and irregularly shaped and have a grayish-brown glazed appearance.

Tomato leaves showing symptoms of septoria leaf spot. Spots have dark borders and light centers with dark specks.

Variety	*Remarks*
Parsnips	
Hollow Crown	Widely adapted
Peas	
Alaska	Resistant to fusarium wilt
Little Marvel	Good quality, good for freezing
Progress No. 9	Large early peas
Peppers	
Calwonder	Sweet, widely adapted
Yolo Wonder	Sweet, resistant to tobacco mosaic virus
Keystone Resistant Giant	Sweet, resistant to tobacco mosaic virus
Hungarian Yellow Wax	Hot
Potato	
Irish Cobbler	Early, widely adapted
Sebago	Widely adapted
Kennebec	Widely adapted
Pumpkin	
Connecticut Field	Large smooth orange fruit
Cinderella	Bush vine, 10-inch fruit
Jack O'Lantern	Large uniform yellow fruit
Radish	
Cherry Belle	Red color
Champion	Red color
Icicle	White color
Southern Peas (cowpeas)	
California Blackeye No. 5	Good for freezing, large seed with black eye
Pink Eye Purple Hull	Elongated seed, light color
Mississippi Silver	Early maturing, large seed, light green to cream in color
Spinach	
Early Hybrid No. 7	Resistant to downy mildew and cucumber mosaic virus
Long Standing Bloomsdale	Widely adapted, slow bolting
Squash (summer)	
Zucchini	Bush plant, cylindrical
Early Prolific Straightneck	Bush plant, tapered cylindrical fruit
Yellow Summer Crookneck	Bush plant, fruit tapered cylinder with curved neck
Early White Bush Scallop	Bush plant, flat round fruit
Squash (winter)	
Waltham Butternut	High quality, elongated pear-shaped fruit
Table Queen	High quality, heart-shaped fruit
Hubbard	Large globular fruit tapered at ends
Golden Delicious	Large, somewhat oval-shaped fruit
Sweetpotato	
Centennial	Widely adapted
Puerto Rico	Widely adapted
Tomato	
Ace VF	Adapted to West, resistant to fusarium and verticillium
Fireball VF	Adapted to East and North, resistant to fusarium and verticillium
Manapal	Widely adapted, resistant to fusarium, gray leaf spot, and blossom end rot
H1350	Adapted East and Midwest, resistant to fusarium and verticillium
Better Boy	Resistant to fusarium, verticillium, and nematodes
Supersonic	Resistant to fusarium and verticillium
Small Fry	Small fruit, resistant to fusarium, verticillium, and nematodes
Turnips	
Purple Top White Globe	Widely adapted
Shogoin	Quick growing, primarily for greens
Watermelon	
Sugar Baby	Small round, early
Crimson Sweet	Blocky oval, resistant to fusarium and anthracnose
Charleston Gray	Oblong, resistant to fusarium and anthracnose

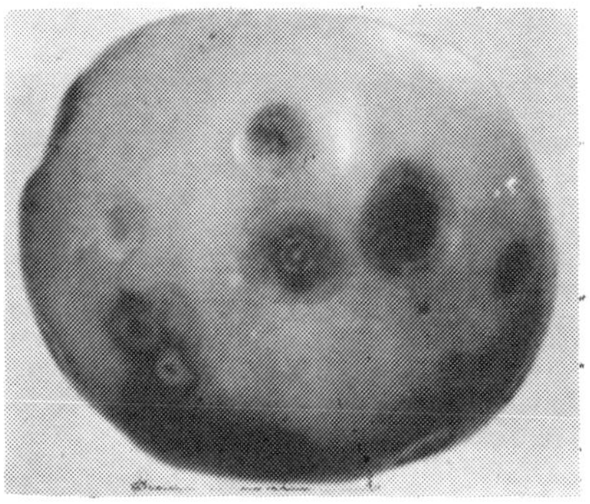

Tomato showing symptoms of anthracnose. Spots are sunken and have targetlike markings.

Tomato showing a late stage of bacterial spot. The spots are raised, but the centers are slightly sunken and have a rough, scabby appearance.

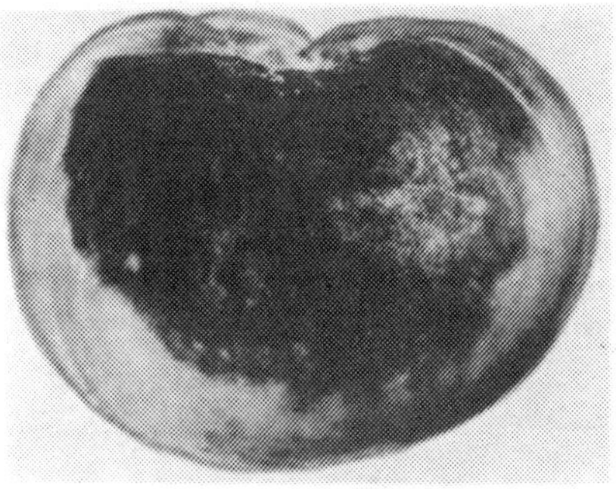

Green tomato affected with late blight. Note wrinkled tissue

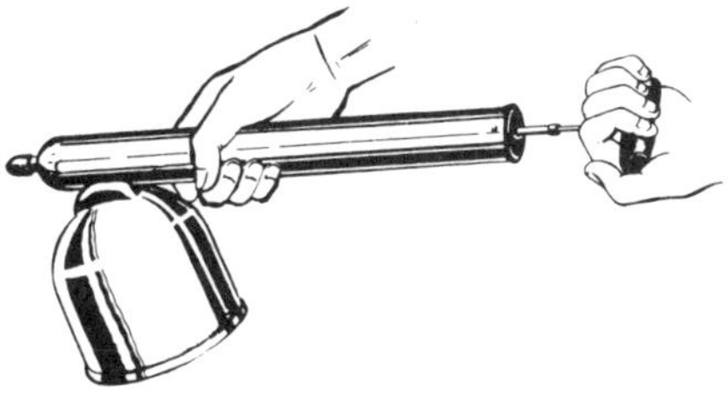

Tomato mosaic

Downy mildew of cucumber

Bacterial wilt of pea.

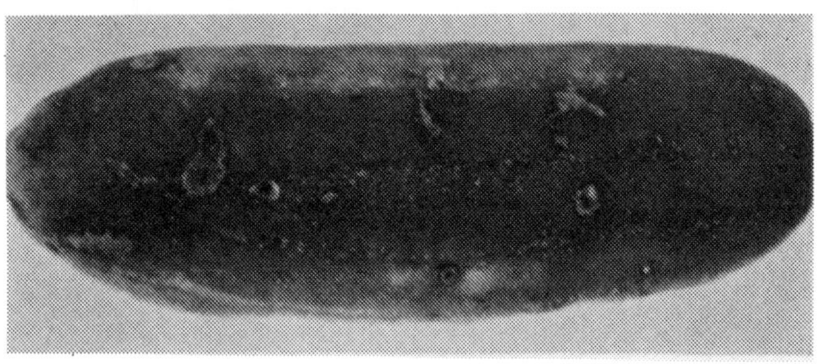
Cucumber scab

Examples of Plant Viruses and Some of Their Insect Vectors

Virus	*Vector*	*Common name*
Potato spindle tuber	*Melanoplus* spp	grasshoppers.
	Epitrix cucumeris	potato flea beetle.
	Systena taeniata	flea beetle.
	Disonycha triangularis	leaf beetle.
	Leptinotarsa decemlineata	Colorado potato beetle.
	Lygus oblineatus	tarnished plant bug.
	Myzus persicae	green peach aphid.
Strawberry yellow edge	*Pentatrichopus fragariae*	aphid.
Strawberry crinkle	*Pentatrichopus fragariae*	aphid.
Onion yellow dwarf	*Aphis gossypii*	melon aphid.
	Myzus persicae	green peach aphid.
	Brevicoryne brassicae	cabbage aphid.
	Aphis maidis	corn leaf aphid.
	Other Aphidae	At least 50 species of aphids transmit this virus.
Cucumber mosaic	*Aphis gossypii*	melon aphid.
	Myzus persicae	green peach aphid.
	Myzus circumflexus	crescent-marked lily aphid.
	Myzus solani	foxglove aphid.
Raspberry mosaics	*Amphorophora rubi*	aphid.
	Amphorophora sensoriata	aphid.
Pea mosaic	*Macrosiphum pisi*	pea aphid.
	Myzus persicae	green peach aphid.
Potato leaf roll	*Myzus persicae*	green peach aphid.
	Myzus circumflexus	crescent-marked lily aphid.
	Myzus solani	foxglove aphid.
	Macrosiphum solanifolii	potato aphid.
Sugarcane mosaic	*Aphis maidis*	corn leaf aphid.
	Hysteroneura setariae	rusty plum aphid.
Citrus quick decline	*Aphis gossypii*	melon aphid.
Potato yellow dwarf	*Aceratagallia sanguinolenta*	clover leafhopper.
	Aceratagallia curvata	leafhopper.
	Aceratagallia longula	leafhopper.
	Aceratagallia obscura	leafhopper.
Sugar-beet curly top	*Circulifer tenellus*	beet leafhopper.
Pierce's disease of grapevines	*Draeculacephala minerva*	leafhopper.
	Helochara delta	leafhopper.
	Carneocephala fulgida	leafhopper.
	Other Cicadellidae	At least 14 species can transmit this virus.
	Aphrophora annulata	spittlebug.
	Aphrophora permutata	spittlebug.
	Clastoptera brunnea	spittlebug.
	Philaenus leucophthalmus	meadow spittlebug.
Phloem necrosis of elm	*Scaphoideus luteolus*	leafhopper.
Peach yellows	*Macropsis trimaculata*	plum leafhopper.
Phony peach	*Homalodisca triquetra*	leafhopper.
	Oncometopia undata	leafhopper.
	Graphocephala versuta	leafhopper.
	Cuerna costalis	leafhopper.
Western X-disease of peach	*Colladonus geminatus*	geminate leafhopper.
Papaya bunchy-top	*Empoasca papayae*	leafhopper.
Cranberry false-blossom	*Scleroracus vaccinii*	blunt-nosed cranberry leafhopper.
Blueberry stunt disease	*Scaphytopius* sp	leafhopper.
Tomato spotted wilt	*Thysanoptera*	thrips.
Tobacco mosaic	Aphidae	A few aphids have been reported to transmit this virus.
	Melanoplus differentialis	differential grasshopper.
Latent potato virus (potato virus X).	*Melanoplus differentialis*	differential grasshopper.
Tobacco ringspot	*Melanoplus differentialis*	differential grasshopper.
Cotton leaf curl (in Africa)	*Bemisia gossypiperda*	whitefly.

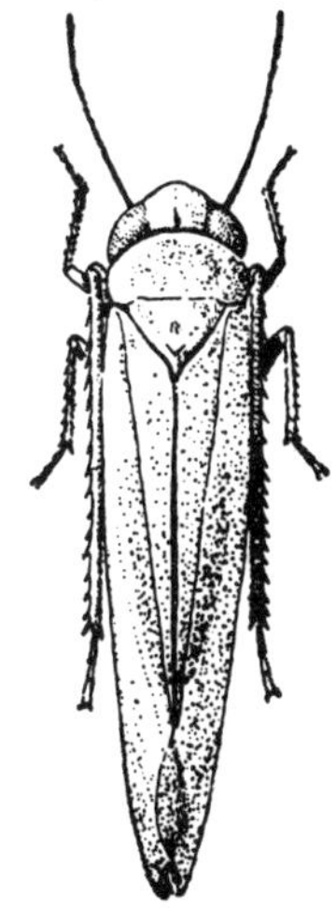

Leafhopper.

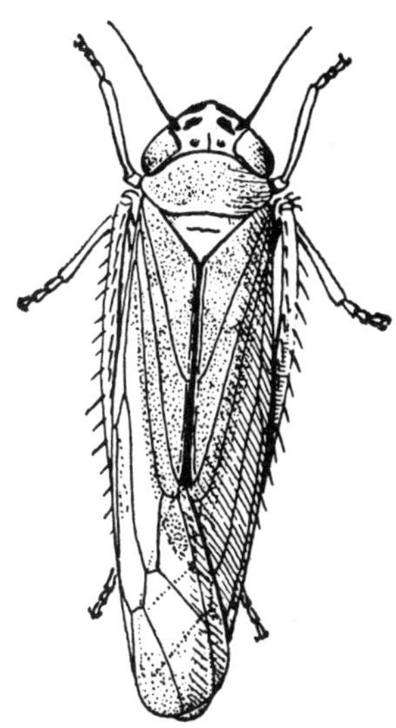

Six-spotted leafhopper.

Seed-corn maggot.

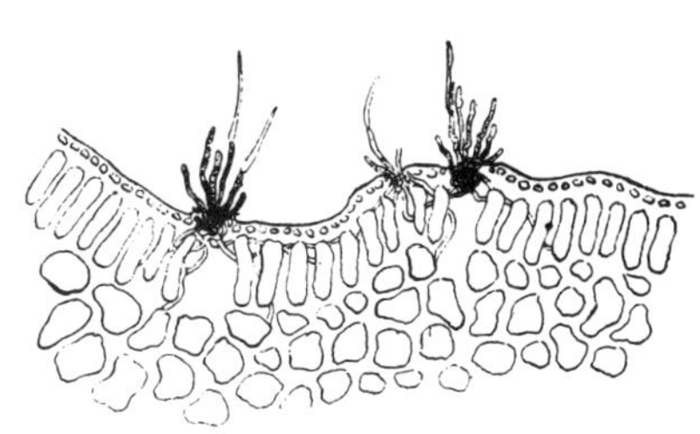

PESTS OF FRUIT TREES, SMALL FRUIT BRAMBLES AND GRAPES

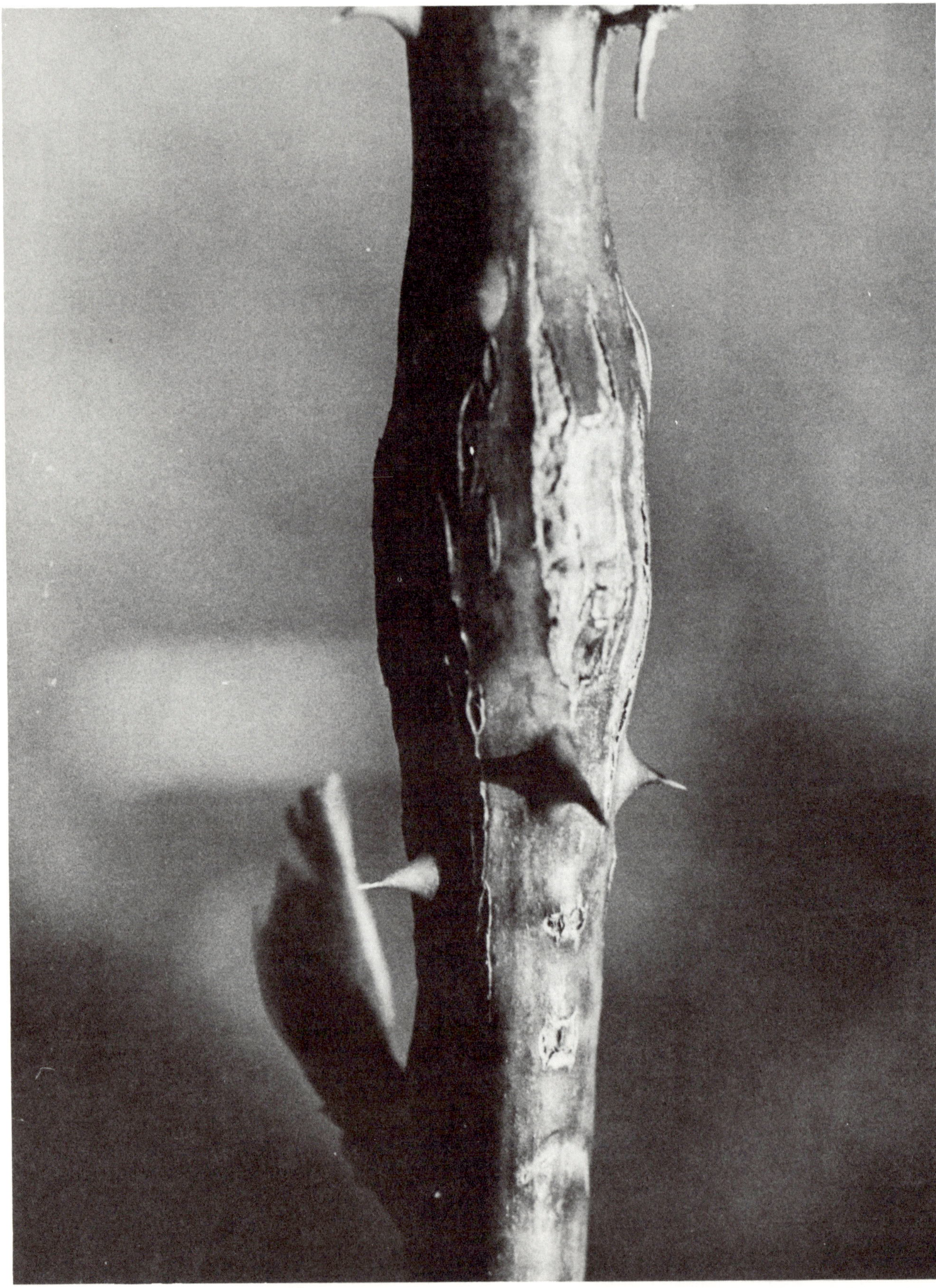

PESTS OF FRUIT TREES

The drawings and charts in this section will deal with the major pests of fruit trees and small fruit. These are the well known insects and diseases that seem to be the most troublesome to home gardeners. It will cover the following trees and shrubs:

APPLES, CRABAPPLES, QUINCE AND PEAR

PEACHES AND NECTARINES

APRICOT AND PLUMS

SWEET AND SOUR CHERRIES

STRAWBERRIES

BLACKBERRIES, RASPBERRIES AND BLUEBERRIES

GRAPES

INSECTS OF FRUIT TREES

APPLE APHIDS

These tiny sucking insects feed on the new terminal buds and fruiting clusters early in the spring, causing fruit to become stunted and knotted, and the leaves to curl and distort. There are several species of apple or fruit aphids and all have the same characteristics of the common aphid. You will probably see the damage and as soon as you do, remove the foliage that has been completely destroyed or spray with an insecticide. They are easy to control once discovered. The green colored apple seems to attract the apple aphid more than the brighter colored varieties.

APPLE MAGGOT

As the name suggests, this insect does a lot of damage to the fruit of apples, plums, crabapples, and blueberries. In some parts of the United States, the apple maggot is called a railroadworm. The legless white maggot tunnels into the apple after the adult fly lays its eggs on the fruit. After an infested apple is picked or falls to the ground, it becomes a brownish, pulpy mass. The flies usually prefer apples that ripen during the summer months and have an acid or sweet taste. The insect over-winters in the pupal stage in the soil. The illustration on the opposite page fully explains the cycles and nature of the infestation.

CODLING MOTH OR APPLEWORM

This is the little whitish-gray or pink worm that is found in apples everywhere in the continental United States. It causes the worm holes on the blossom end of the apple. The entry holes are usually covered with a dark mass or with pellets that project out of the hole. This insect also attacks English walnuts, pears, and quince. The worms over-winter in cocoons in broken parts of the bark and any protected area on or near the fruit tree. The moths emerge about the time the apple tree comes into flower and stay around the greater part of the growing season. The tiny white eggs are laid on the foliage next to the fruit or flower, and sometimes right on the fruit. The worms begin to enter the fruit a few weeks after the flowers have dropped.

GREEN FRUITWORMS

This insect eats irregular holes in the new growth in the spring. As the plant matures the worm tunnels into the fruit. The worms rest on small twigs and can be hand picked.

ORIENTAL FRUIT MOTH

This insect damages peaches, and nectarines. It does limited damage to quince but is not really considered a serious problem. In the larva stage the worm enters the fruit and young tender twigs. The illustration shows the various stages of the insect and the damage. The worms feed first on the twigs, boring into the tender parts of the shoots, thus causing them to wilt. As the stems harden, the worms turn their attention toward the fruit, and bore into it. They over-winter in the cocoon stage in a protected place on the plant or nearby on the ground. In some of the warmer parts of the country you can expect a few generations per season.

ORIENTAL FRUIT MOTH

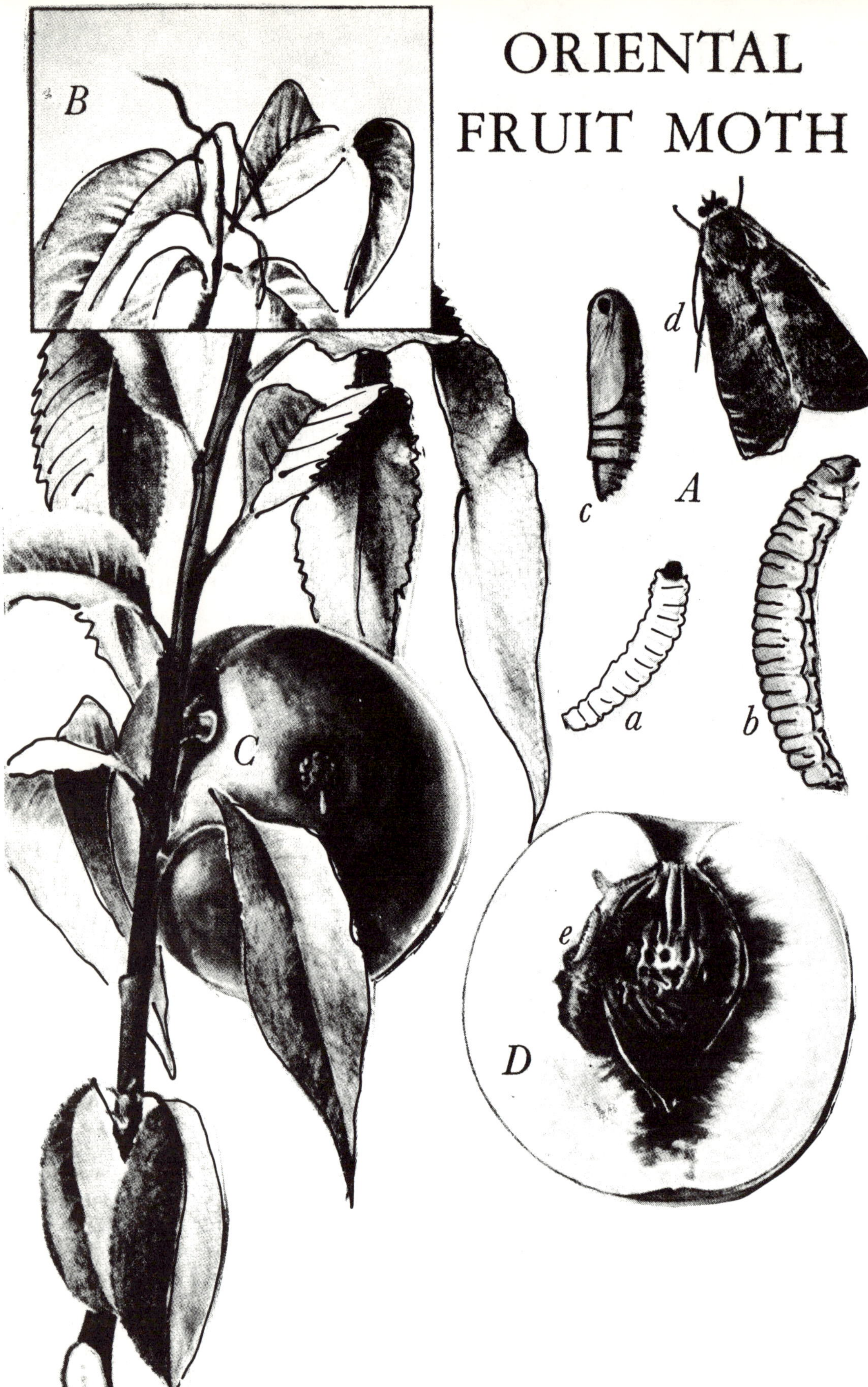

A. Life stages; a. young larva; b. mature larva; c. pupa; d. adul
B. Damage to tender new tips
C. External evidence of fruit moth damage to ripe peach.
D. External damage with e, larva and frass next to stone
(all greatly enlarged).

APPLE MAGGOT

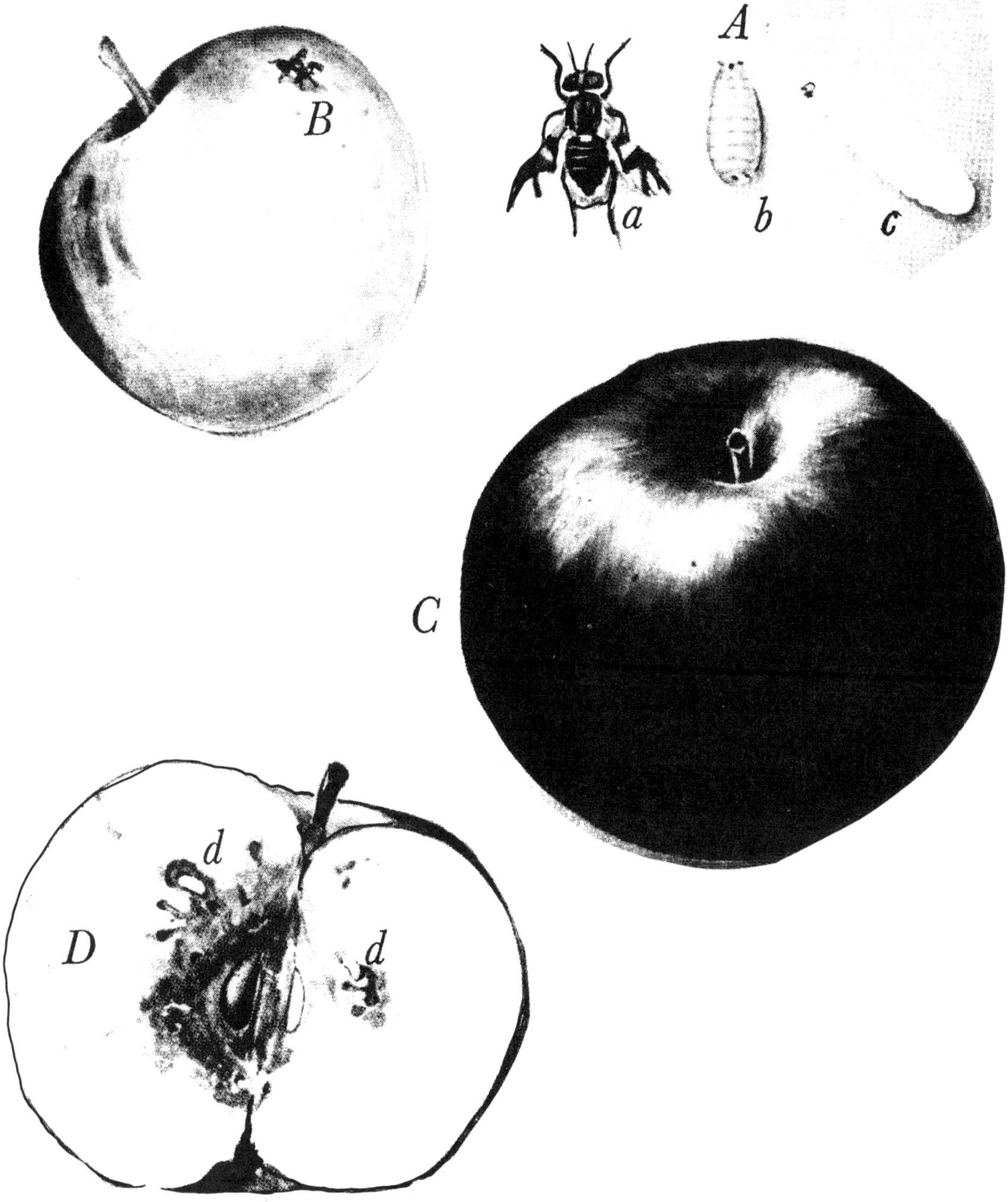

Apple maggot

A. Life stages; a. adult; b. pupa; c. larva, or maggot
B. Adult female making egg punctures in an apple.
C. Exterior evidence of apple maggot damage.
D. Halved apple showing damage to apple flesh and d, maggots feeding inside. (All greatly enlarged.)

PEACH TREE BORER

This insect attacks both the stone fruit, (peaches and cherries) and the woody parts of the plant. The peach tree borer usually develops and feeds just above or below the soil line on the woody stem. A similar borer (the Lesser Peach Tree Borer), may attack anywhere on the tree or shrub. The adult moth lays the eggs on very rough surfaces of the bark. The hatchings must get under the bark for protection; which causes damage to the woody stems and the bark. If not controlled as soon as you notice the borings (small holes) with wood dust or gum masses extruding from the plant, this insect can girdle the plant and kill it. Both of these insects over-winter on the bark in the larva stage. See drawing for description and damage.

PLUM CURCULIO

This is a serious pest of apples, plums, cherries, and peaches in all sections of the country West of the Rockies. The fruit is marred on the surface and generally distorted by the feeding and egg-laying habits of the insect. The illustration clearly indicates the extent of the damage and the appearance of the pest. The worst damage is done by the larva that tunnels into the fruit.

The adult plum curculio is a brown beetle with a snout and a humpback, ranging from one fourth to one third of an inch. They feed first on the tender foliage, move onto the flowers, then after the fruit develops onto the fruit, puncture the soft tissue, and lay their eggs. The eggs are laid in moon shaped slits on the fruit (see drawing), causing scabs to form. The yellowish-white grubs begin hatching in about two weeks and feed on the fruit. As they reach maturity, they leave the fruit and drop to the soil, thus completing the life cycle. In the South there may be two or more generations in one growing season. In the North, say North of the Delaware watershed region, you'll only find one generation. You can use a dormant spray in the late fall or early spring before the plants develop buds or after the plant has gone dormant.

RED-BANDED LEAF ROLLER

The adult pest feeds on the foliage and the larva feeds on the surface of the apple, pear, and quince, producing a scar and sometimes a depression in the fruit. The first damage is done in the early part of the growing season. This attack is centered on the new green growth and newly-formed fruit (just after the flowers drop). The second attack is directly on the mature fruit causing very heavy damage in the late summer just before harvest. This insect over-winters in rolled leaves on the ground. If you cultivate around the trees, not digging too deeply, around new plant stock, you can kill the pest in this stage without using any chemicals.

Lesser peach tree borer larva and injury.

Plum curculio adult and egg puncture.

PLUM CURCULIO

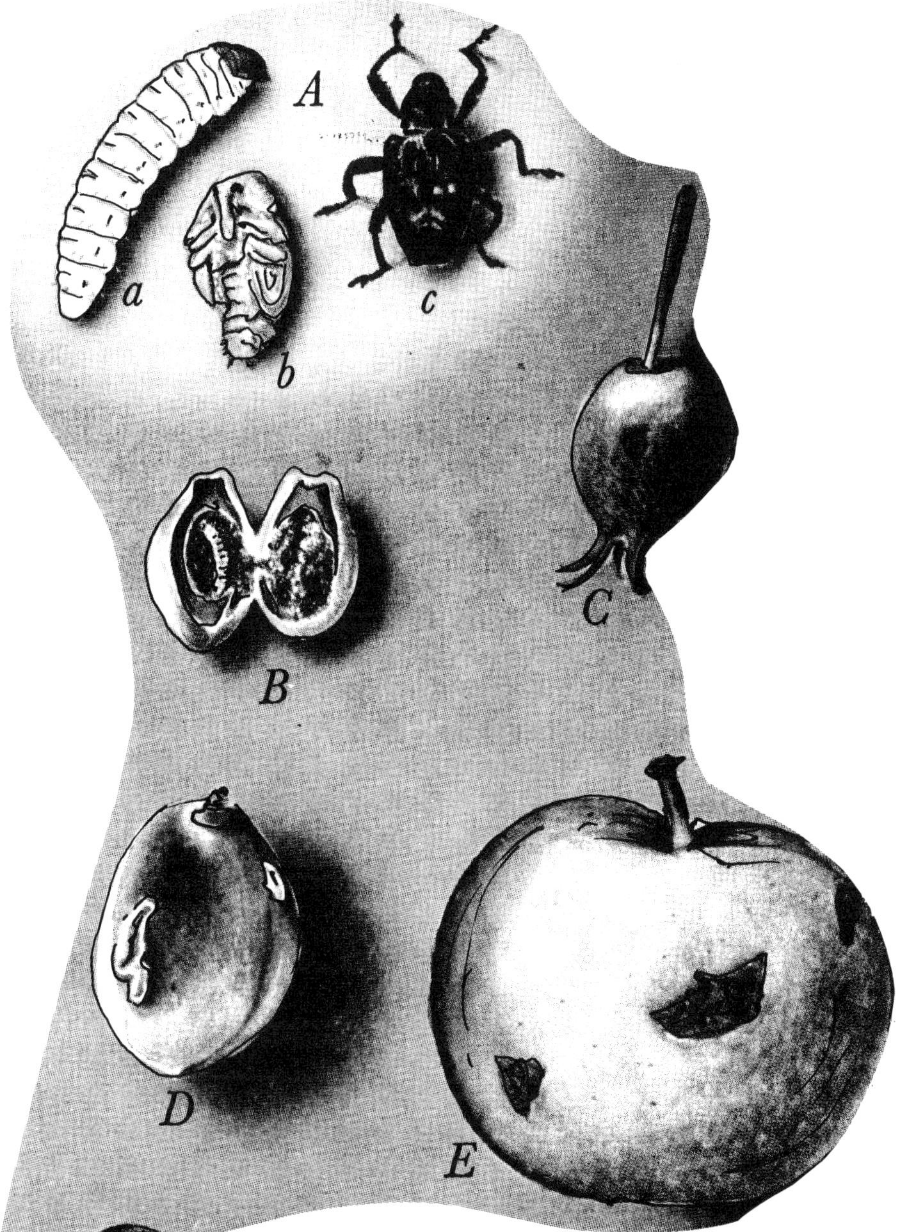

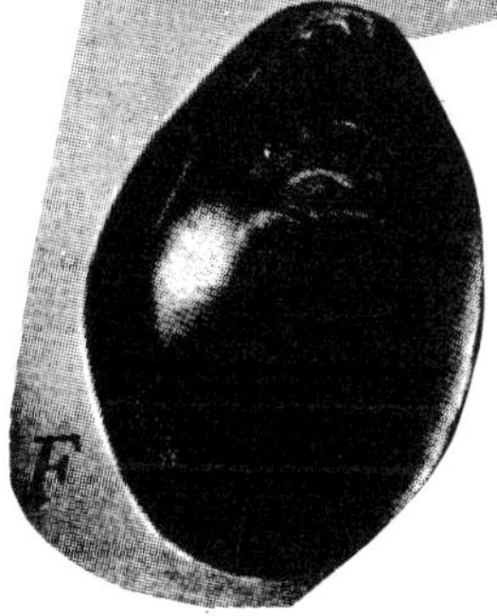

Plum Curculio

A. Life stages; a. larva; b. pupa; c. adult.
B. Peach "mummy" with larva inside.
C. Small apple showing curculio scar. D. Small peach with gum oozing from curculio puncture.
E and F, mature apple and plum showing appearance of curculio scars at time of harvest. (All greatly enlarged.)

CODLING MOTH

A. Adult moth B. Twig and fruits showing a. eggs on the underside of leaves; b. larva and internal damage; c. sting; d. surface breakdown and frass from internal feeding; e. frass-plugged emergence hole. C. Section of bark from trunk or large branch showing f. hibernating larva and g. pupal skin from which adult has emerged.

SAN JOSE SCALE, FORBES SCALE

This is a tiny insect that sucks the sap from the woody stems, the foliage and, the fruit of the apple, peach, pear, and other fruit trees. The mature scale is the size of a pin head, and can only be seen in clusters on the stems and leaves. The San Jose scale is yellowish in color but when a heavy infestation appears on the plant it looks reddish. Fruit trees that are heavily infested take on a rough appearance. The tiny new-born are called crawlers. They remain in one place during the first few hours after hatching, then move on the host plant. At times they are blown to other plants. These insects reproduce very fast and can produce many generations in one season.

In the mid-West, Forbes scale greatly resembles the San Jose scale. The appearance is about the same and the seasonal cycles are the same. The drawing of San Jose scale explains the life cycle and how it usually looks on the plant. Leaves with a heavy infestation usually drop from the plant and the plant loses vigor. The fruit has a spotted appearance.

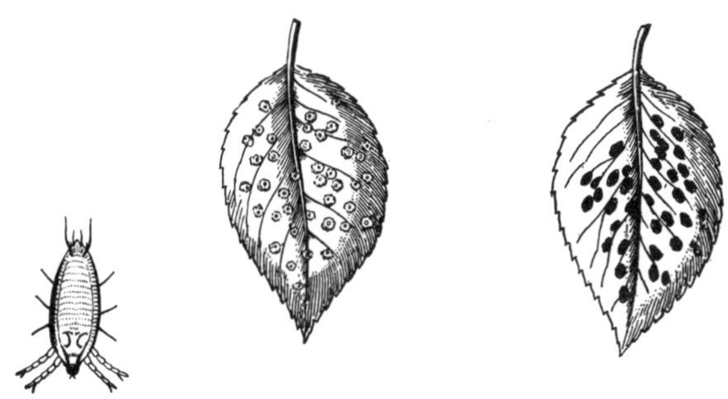

TWO-SPOTTED SPIDER MITE

Sometimes called orchard mites, these tiny sucking insects are dreaded by the commercial fruit grower. They attack apples, cherries, pears, peaches, plums, and other deciduous fruit. The European red mite fits into the same category as the two-spotted spider mite. Most mites have the same life cycle and the webs they "spin" (thus spider mites), can be seen before you see the tiny insect-this is sometimes the only way you can detect their presence. They infest leaves and tender stems. The clover mite, that attacks some fruit trees, and the European spider mite and the Pacific mite over-winter as the adult females in protected places on the ground. The complete cycle from egg to adult takes roughly ten days. If the weather is hot and dry, the development is faster and may favor a serious outbreak of mite trouble on your fruit trees. Anyone that has grown small fruit knows the worst damage is done during the hottest months of the growing season.

Control of mites on trees is difficult but a dormant spray in the early spring before the buds appear, or late in the fall after the tree has dropped all of its foliage is probably the best method. You can use a miticide chemical spray but I recommend a dormant spray for the best results of all sucking insects.

A dormant spray is an oil preparation that covers the branches and twigs of the tree and prevents the insects from developing. It is used by all commercial orchards as a control against mites, scale, and any insect that over-winters on the tree. As soon as the first heat of spring comes, the oil just sort of fades away. Never spray a plant with green foliage or new buds with dormant spray. It will damage the plant and kill the buds.

TWO-SPOTTED SPIDER MITE
EUROPEAN RED MITE

A. Peach tree twig showing typical russeting of foliage of mites.
B. Life stages of the two-spotted spider mite; a. egg; b. young (note, six-legged at this stage); c. half grown; d. adult male; e. adult female; and f. overwintering female.
C. Russeting of apple foliage.
D. Stages of European red mite; g. egg; h and i. nymphs; and k. adult female. (all greatly enlarged.)

SAN JOSE SCALE

A and B. Infestation on fruit
C. Typical damage to new growth
D. A small cluster of scales (greatly enlarged)

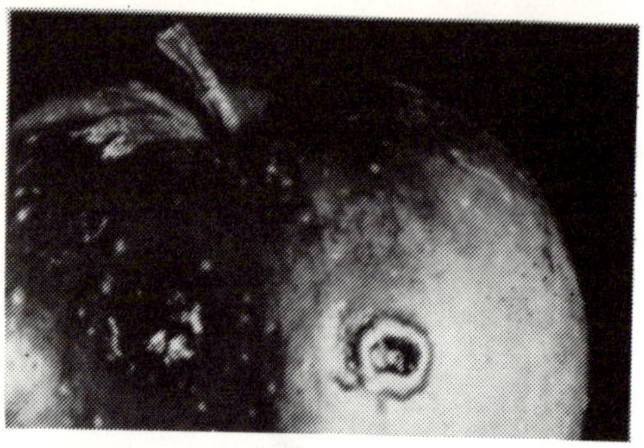

Codling moth wormholes in apple

Leafroller larva damaging apple

INSECTS

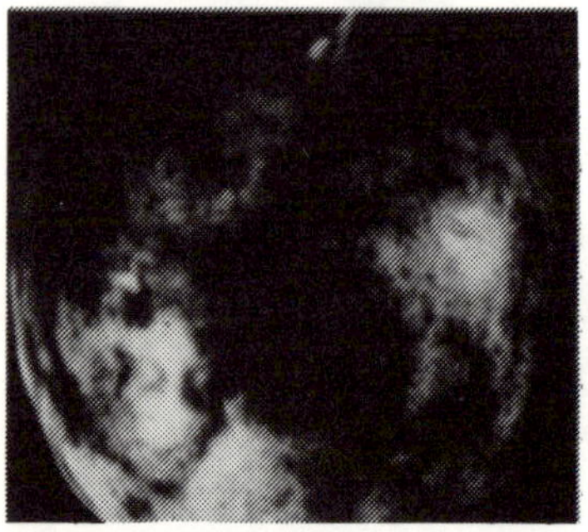

Injury by apple maggot

Aphid injury to aphid leaf

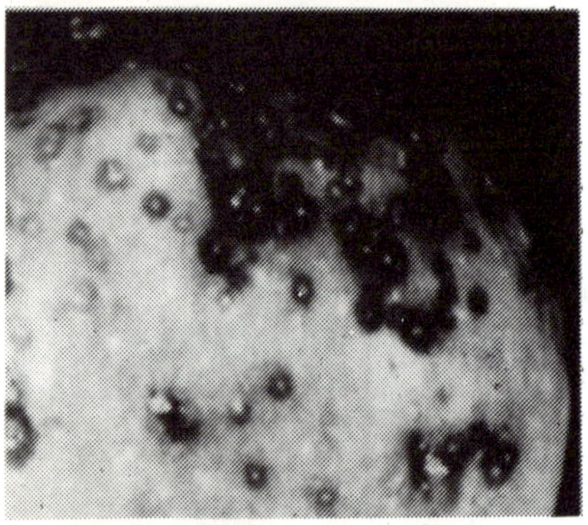

San Jose scale on Golden Delicious apple

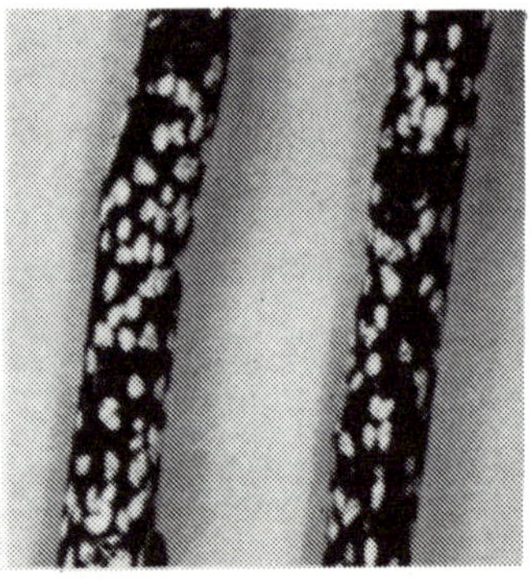

Scurfy scale on apple twig

Periodial cicada adult on apple limb

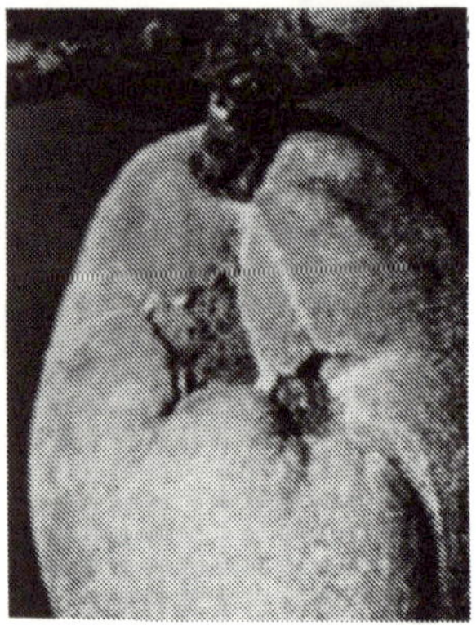

Plant bug injury (catfacing)

Spider mite on apple plant

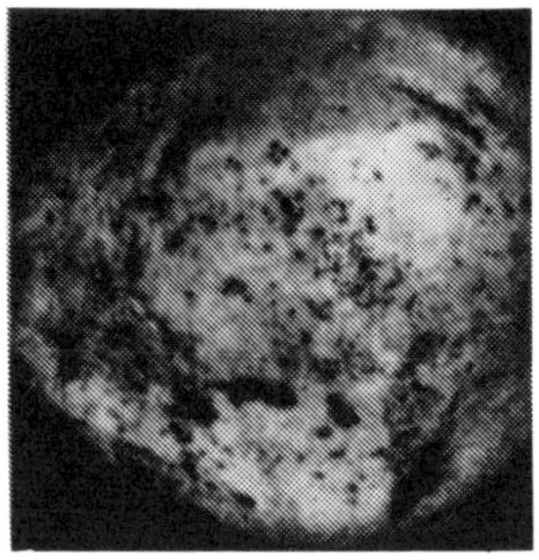

Pear psylla damage

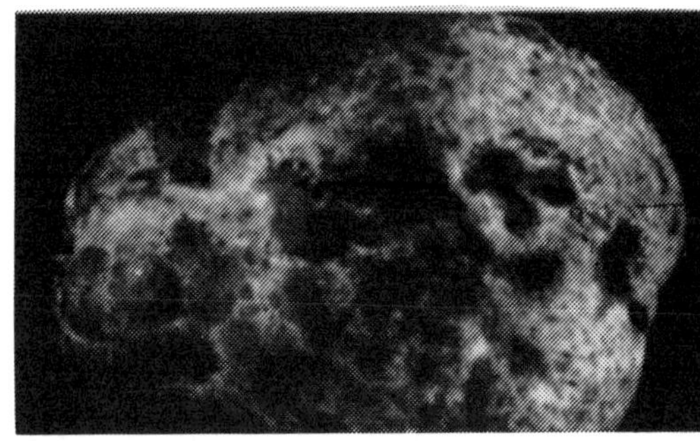

Pear leaf blister mite damage

Cherry fruit fly maggots and damage

Walnut caterpillar larvae feeding on leaflet

Casebearer larva and case

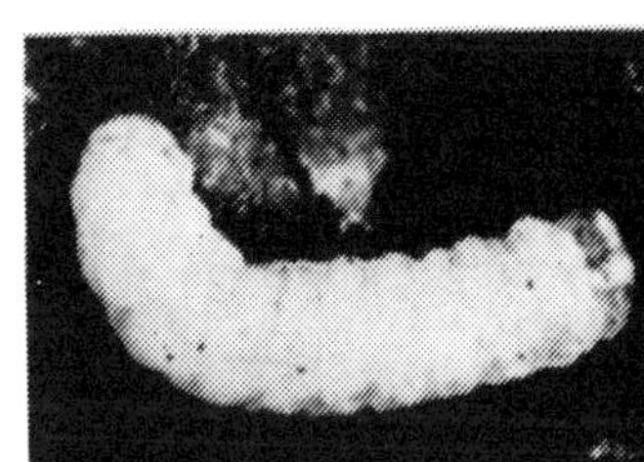

Hickory shuckworm larva

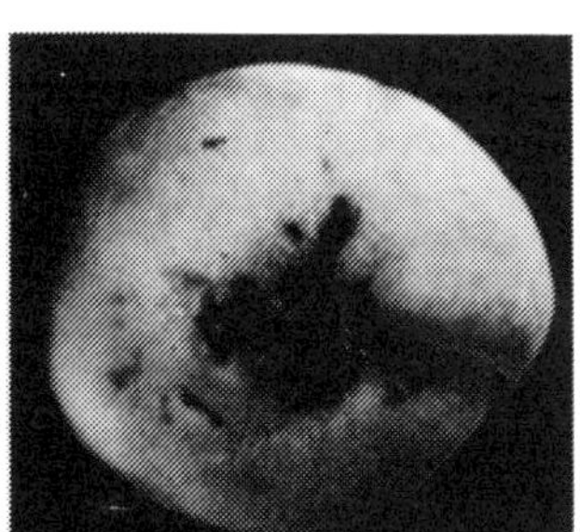

Oriental fruit moth injury to peach

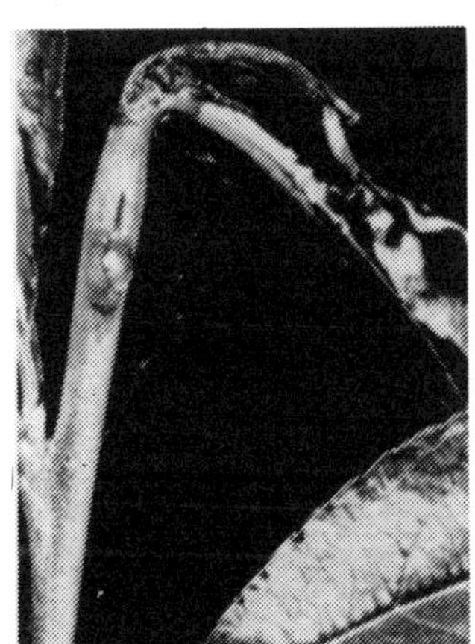

Oriental fruit moth damage to peach twig

Pecan weevil adult

If you use a miticide, do not spray for three weeks before harvesting the fruit. If you insist in using chemicals for control, peel the fruit before eating. I always wash fruit before eating, but if I know it has been sprayed with a chemical, I am more careful than usual. Make sure that all vegetables or fruit do not have a chemical residue on them before freezing or canning, or just eating.

I know today's fruit, as seen in the supermarkets, are nice and red or yellow, without a mark on them. But a few scars or holes, a little off-color, will not hurt you-fruit does not have to look like pictures in magazines. Apple scab is as natural as the insects that live in the apple tree. Try and live with real fruit and do not expect the supermarket colors, for most of the fruit has been colored artificially for your benefit. It's better to eat healthy fruit than try and match a color photograph that has been air brushed by a commercial artist.

APPLE SCAB

This is a widespread disease of standard and dwarf apple trees as well as many varieties of crabapples. The foliage develops a dull smoky spot and becomes an olive green color, with the texture of velvet. If the disease becomes serious, the leaves may drop from the plant or scab and develop a wilt. The fruit develops a black spot with scaly formations. If the disease is not controlled and the plant is affected early in the spring, the fruit may become deformed and cracked and drop from the tree.

BLOTCH

This disease infects all parts of the fruit tree; the leaves, the stems, and the fruit. The spots are small, light gray with a black dot in the middle, or many times the spots are elongated, light colored, and with several black dots.

BROWN ROT

This is the most destructive disease of ripe peaches, plums, nectarines, cherries, and apricots. The disease areas are soft and brown and rapidly enlarge as the disease spreads. Tufts of gray mold in circles cover the rotten spots. The fruit that is completely infested will shrivel up and become hard. Beginning symptoms are blighting of the flowers and the new green growth. Stems and young twigs may become brown in color and die back.

CEDAR-APPLE RUST

In late spring around May or early June, small light yellow spots form on the leaf surfaces. As the growing season continues, the spots develop into much larger spots. Small black specks later form in the center of the orange circles. Rust spots may appear on the fruit near the blossom end. If the fruit is infested early in the season, it probably will drop off the tree.

The quince rust is dark green with crater-like spots on the fruit, which then becomes soft and brown. This does not affect the other parts of the plant. The upright junipers and true cedars are the host plant of this disease. Fruit trees should never be planted in areas that have these host plants, as they disperse spores. Understanding the host plant and the possibility of infestation is a natural control by itself.

CHERRY-LEAF SPOT (YELLOW LEAF)

Many tiny reddish-blue spots form on the foliage in early summer. The spots enlarge by midsummer, turn brown, and eventually drop from the tree. It looks as if a shotgun has been fired into the foliage with a fine birdshot. At this point the leaves are completely yellow and if the infestation is widespread, the fruit will not develop properly. This disease can defoliate a plant in one season, and if this happens the plant can die.

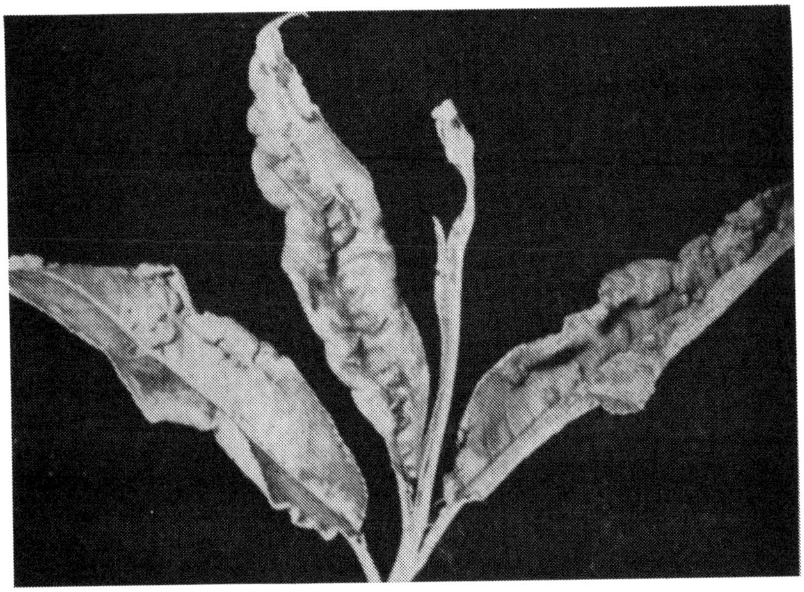

Peach leaves distorted by peach leaf curl disease. The distorted parts usually turn reddish-purple in color. Badly infected leaves drop in early summer.

Mildew on apple

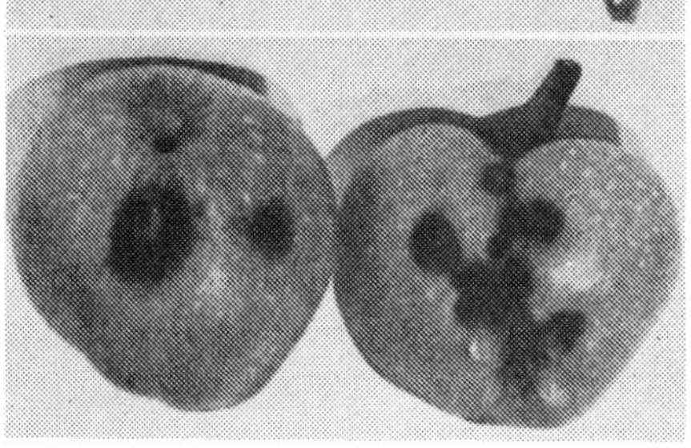

Apple scab

Cedar-apple rust on apple leaves. These lesions are orange in color with a small black center.

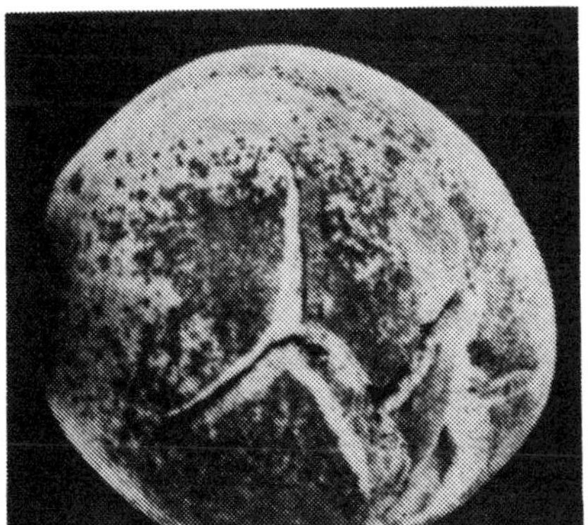

Peach scab

Cherry leaf spot

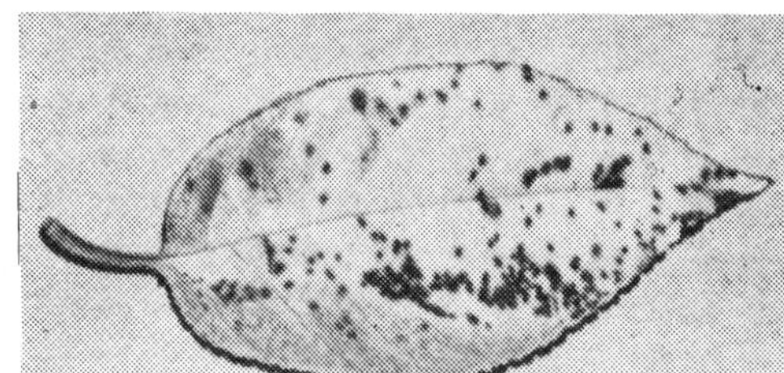

FIRE-BLIGHT

This disease is a disaster. It is a common problem that attacks not only a number of fruit trees (apples, crabapples, and pears), but also many ornamentals. It is listed in the fruit section for it can wipe out a plant in one season, and many times you do not see it coming. You may observe fire blight before it hits your fruit tree, if you have any of these ornamentals in the area: quince, mountain ash, firethorn (pyracantha), spirea, serviceberry (amelanchier), and rose. This bacterial disease will kill all of the above as well as your apple and pear trees. If the plant is attacked in the early spring the young spurs and the flowers will wither and die. During the hot months of the growing season, once the disease has gotten a firm grip on the plant, the foliage looks as if it has been burned. This bacterial disease over-winters in cankers on the woody parts of the plant. The cankers are sunken and discolored with a sharp margin.

PEACH-LEAF CURL

This fungus causes blistering of the foliage and curling of the leaves. The leaves turn a dark red color (almost purple) and drop off the tree. The drawings and photographs show the general characteristics of this fungus.

Plum Pockets or blatter plum is another fungus that is a close relative of peach leaf curl. The plums become large, light colored, become wrinkled, and will drop from the tree.

PEACH SCAB

Dark green to black spots form on the fruit. If the disease is severe, the spots seem to merge and form one dark area. The fruit develops cracks and is irregular in shape.

POWDERY MILDEW

This disease is very easy to see--it is a whitish-gray feltlike patch growing on the foliage and new growth. The leaves fold upward and become dwarfed. The fruit sometimes looks like a brown rusty net and becomes cracked.

PEACH SCAB

Siberian Crab (*Pyrus baccata*).

Sour or pie cherries.

	BLACK WALNUT	BUTTERNUT	CHINESE CHESTNUT	HAZELS AND FILBERTS	HEARTNUT	HICKORIES	PERSIAN WALNUT
Climatic Region	Well adapted to southern Michigan.	Extremely hardy.	Chestnuts are about as hardy as peaches. Planting should be confined to areas where peaches are winter hardy.	The American hazel is fully hardy, but the European hazel or filbert isn't. Hybrids between these two species vary in hardiness.	Extremely hardy.	Shagbark hickories do well throughout lower Michigan, except in extremely dry or wet areas.	Early maturing varieties are satisfactory as far north as M-46 in Michigan and perhaps farther along the lake where the growing season is 150 days or more.
Soil	Does best in deep, fertile, moist, drained soils.	Does best in deep fertile, moist soils.	Well-drained gravelly type of soil, suitable for peaches, is ideal. Soil should be slightly acid. Avoid heavy, strongly alkaline and poorly-drained soils.	Good water drainage is essential.	Will grow on infertile sandy soils, but does better on deep, fertile soils.	Prefer light, well drained, loamy soils.	Needs deep, fertile soil which retains moisture in summer. Swampy, poorly-drained areas and droughty, eroded hillsides are not suitable.
Air Drainage	Avoid frost pockets.	Will do better in frost pockets than black walnuts.	Good air drainage required. Late spring freezes may kill new shoots and tender leaves if planted in a frost pocket.	Good air drainage is essential.	Avoid frost pockets.	Avoid frost pockets.	A fair amount of air drainage is desirable. Trees in frost pockets may have blossoms and new leaves killed by late spring frost.
Site	Yards, fence rows, on northern or eastern slopes, permanent pastures, large lawns.	Yards, fence rows, permanent pastures, large lawns, low hillsides, river banks.	Does best on slopes and hilltops where there is good air drainage.	Do well on moist sites if soil drainage is adequate. Low areas that form frost pockets should be avoided because of both winter cold and spring frost.	Yards, fence rows, large lawns.	Low hillsides and river banks.	Does well in pastures and other grassy areas if sufficient soil moisture and nutrients are available. Level or gently-sloping land is best.
Varieties for Michigan	Adams, Allen, Beck, Elmer Myers, Grundy, Snyder, Sparrow, Stambaugh, Thomas.	Crackezy, Lingle, Love, Sherwood, White.	Abundance, Crane, Kuling, Meiling, Nanking.	Bixby, Buchanan (both American hazels); Italian Red, Latson, Potomac, Reed, Rush, Winkler.	Fodermaier, Mitchell, Walters, Wright.	Abscoda, Glover, Kirtland, Lingenfelter, Romig, Wescheke.	Broadview, Colby, Fickes, Gerstenmaier, Gratiot, Greenhaven, Hansen, Lake, Merkel, Metcalfe, McDermid, McIntyre, McKinster, Neyer No. 1, Somers, Winchell N.
Culture	Weed control is necessary for newly planted and small trees. Mulching with straw or wood shavings is beneficial, particularly during dry periods in the summer. Trees need little or no pruning. An occasional limb may be shortened or removed to maintain symmetry.		Weed control is important, especially with young trees. Mulching should help. Keep sod clipped (4 to 5 times/year.) Extremely low crotches should be avoided by removing limbs to a height of 2 to 4 feet above ground. Scaffold branches may then be selected 18 to 24 feet apart to form a modified leader type of tree. Subsequent pruning will be very light.	Weed control is necessary for newly planted and small trees. Mulching with straw or wood shavings is beneficial, particularly during dry periods in the summer. Trees need little or no pruning. An occasional limb may be shortened or removed to maintain the symmetry of form.			
	Black walnuts are usually pollinated naturally.	Usually pollinated naturally or self pollinated.	Need more than one variety (at least 2 or 3) for pollination.	Need more than one variety (at least 2 or 3) for pollination.	Usually pollinated naturally or self pollinated.		Pollen shedding and stigma receptivity usually occur at different times. Therefore, 3 or more different varieties should be planted to insure proper exchange of pollen.
Time of Bearing	Seedling trees come into production in 8 to 12 years. Grafted trees will start to bear earlier.	Named varieties usually begin bearing 2 to 4 years after planting.	Seedling trees come into bearing at about 7 years of age. Grafted trees start to bear about a year after planting.	Named varieties should start bearing the third year after transplanting, if properly cared for.	Named varieties usually begin bearing 2 to 4 years after planting.	Named varieties usually bear in 10 to 20 years.	Seeding trees come into production in 6 to 10 years. Grafted trees bear earlier.
Yield	Variable. Tend to have alternate bearing habits. One to three bushels of hulled nuts per mature tree.	One to several bushels of hulled nuts per mature tree.	Yields vary, but usually average about 20 to 50 pounds of hulled nuts per mature tree.	One to 6 quarts of hulled nuts per mature bush.	One to 6 bushels of hulled nuts per mature tree.	One half to 1½ bushels of hulled nuts per mature tree.	Mature trees have been known to yield 200 lbs. of hulled nuts per tree (about 6 bushels). Good care is necessary to produce good yields. Average yields of 50 to 125 lbs. can be expected.

PESTS OF NUT TREES

Squirrels and chipmunks probably cause the greatest loss of nuts, but insects and diseases are also pests of nut trees and their fruit. Control is not commonly practiced on a regular schedule.

In black walnuts, walnut husk maggot larvae may infect nut husks. Anthracnose, a disease that is common throughout the vegetable world, is a serious problem with nut trees. It may cause early loss of leaves. Army worms, in periodic outbreaks, may eat all foliage on some branches or may strip an entire tree. A few years of complete defoliation by these pests may weaken the tree and sometimes cause death.

In butternut, a fungus causes dieback. It infects the limbs of weaker trees. The limbs die progressively from twigs to the larger branches.

Hickories are not seriously affected by insects or diseases.

Hazelnuts are often destroyed by a fruit worm.

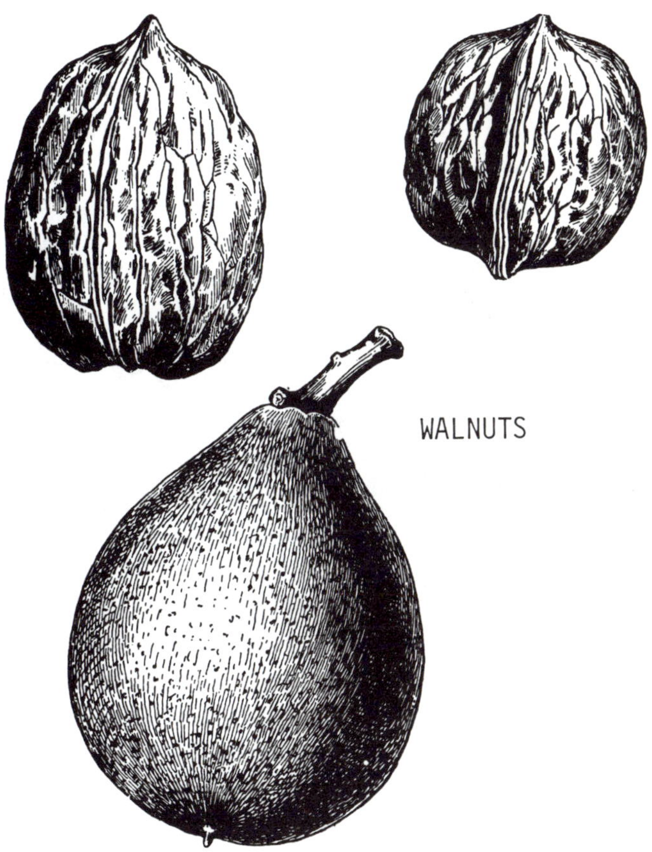

WALNUTS

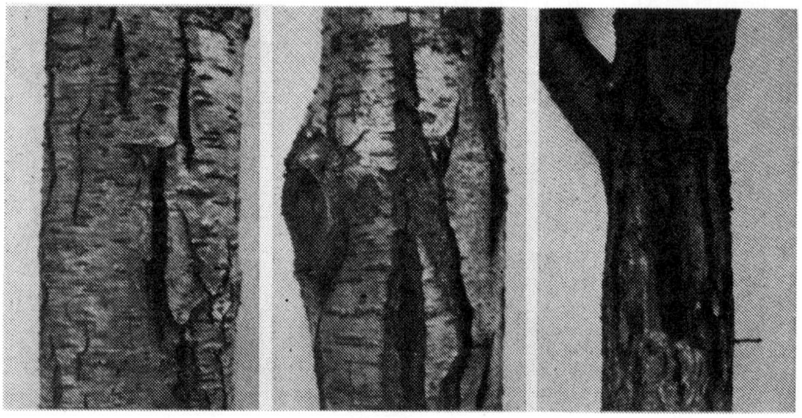

Winter injury to trunks of plum trees on the left and in the center and to the trunk of the peach tree on the right.

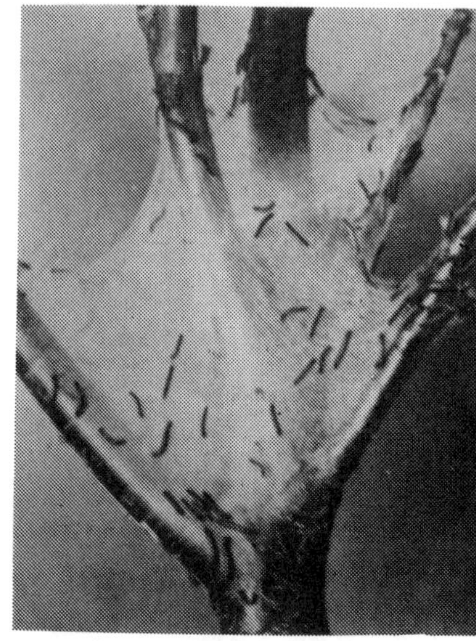

Nests of the eastern tent caterpillar: note the layers of silk.

Character of larval mines of oak-destroying bark-borer in inner suface of bark.

Larval mines of destruction bark-borer in surface of dying cottonwood tree.

WINTER INJURY OF FRUIT TREES

CAUSES

1. Extremely low temperatures.

2. Temperature changes from low to above normal over a few days time span.

3. A cold frost or extreme temperature below freezing in the early fall before the plant has had a chance to harden off (adjust to the temperature change in a gradual manner).

4. Early spring frosts: If the plant has started to develop green growth and a hard frost hits, the new terminal buds are usually killed back.

5. Extreme fluctuation in day and night temperatures.

6. Winter sun damage: The heat of the winter sun on one side of the trunk and cold temperatures on the other side, causes the bark to crack and split. This is the perfect opening for insects to enter the tree trunk in the spring.

7. Ice or heavy snow on the foliage or broadleaf evergreens or branches of deciduous and evergreen trees. In the winter of 1977-78 the East suffered from one of the worst ice storms in history, with extensive damage on the Eastern Seaboard (see photographs).

The photographs indicate the amount of damage that can be caused by winter temperature changes. It is difficult to prevent splitting on large trees (many times splitting will not do a great deal of damage to a large tree), but small, newly planted fruit trees should be wrapped with a "tree wrap". I would also stake new trees to prevent roots moving in the wind.

PREVENTIVE CARE

The best way to avoid injury that is caused by winter temperature changes or early cold spells is cultural.

Avoid any fertilizing with a nitrogen fertilizer late in the season. You want the plant to stop growth as soon as possible after the harvest of the fruit.

Do not water the plant in excess during the late part of the season.

Avoid late cultivation in the orchard. These simple practices will help the plant prepare for the winter months. A little trick used by a lot of gardeners is to paint the tree wrap covering with a white paint to reflect the sunlight, thus helping prevent cold and hot temperatures on the trunk of the tree.

PESTS OF BLACKBERRIES

Extreme damage by insects is not common but mites, thrips, tarnished bugs, and the adult beetle of the raspberry worm do feed on the flower buds and the interior of the flowering parts (stamens and pistils). Good cultural controls including thinning of the plants and keeping the plants off the ground by stalking, will prevent insect infestations.

FUNGUS INFECTION

Anthracnose is the most common disease that affects the blackberry plant. Trailing blackberries seem to be more susceptible than the upright varieties. To prevent most fungus attacks on blackberries, remove the fruiting canes soon after harvest. This will usually eliminate all the fungus activity of that year and give the new growth in the spring a clean start.

VERTICILLIUM WILT (See Raspberries)

VIRUS INFECTION

You cannot successfully use chemicals against virus problems in bramble fruit but you can follow these simple cultural procedures.

1. Thin out plants that do not bear fruit or set fruit. Take out roots and all without disturbing the other nearby canes.

2. Never use root suckers to propagate new canes that come from problem areas.

3. Remove all canes after harvest by cutting them back.

4. All uncultivated brambles in the area should be removed. This is a breeding ground for many fungus and virus related problems.

5. Buy good stock from established growers.

Signs of virus attacks on blackberries are the following:

1. Affected plants seem to be more vigorous with rounder and glossier foliage.

2. The leaves develop a reddish color far in advance of the normal fall period.

3. The flowers seem to be normal but only a few drupelets will develop. This virus will spread by means yet to be discovered.

Black raspberries before and after dormant pruning. Purple raspberries are prune the same way, except the side branches are cut back to 24 inches.

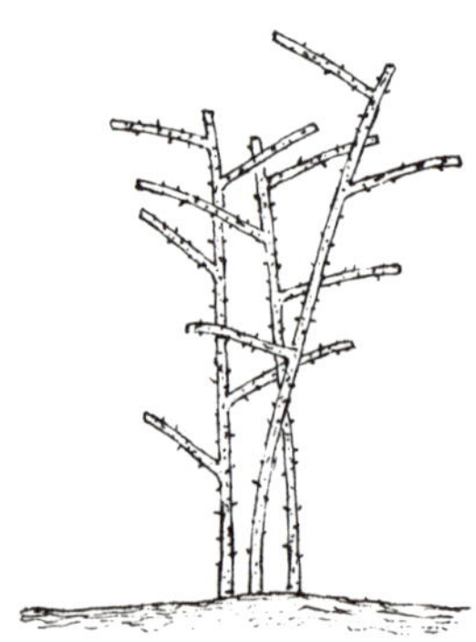

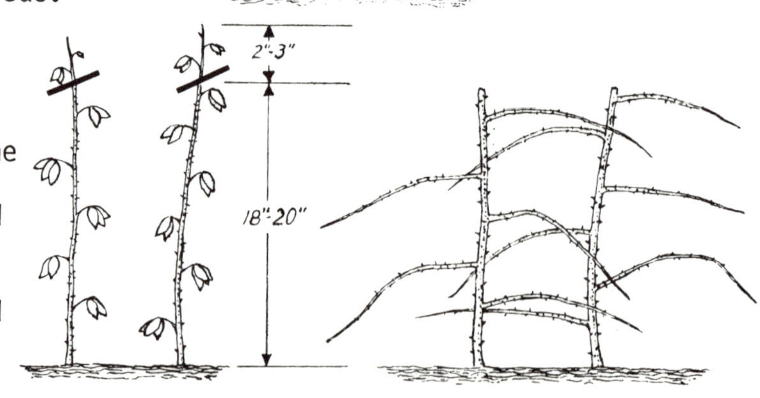

Tipping or summer pruning. Both black and purple raspberries are "tipped". Purple raspberries are more vigorous than the blacks, so let them grow to 30 inches before tipping.

PESTS OF RASPBERRIES

RASPBERRY CANE MAGGOT

The adult is a small gray fly that lays its eggs on the new buds or in the tips of the shoots. As the eggs hatch the maggots begin feeding in the tips and buds, completely girdling the cane. The maggots then eat down the stem and remain all winter. This is not a serious problem for the home gardener for the insects only attack a few canes. If you have only a few canes it could spell the end of the raspberry jam you were planning on. If you can spot the infestation, merely prune out the bad canes. There is no reason ever to use chemicals in insect control in small garden plots.

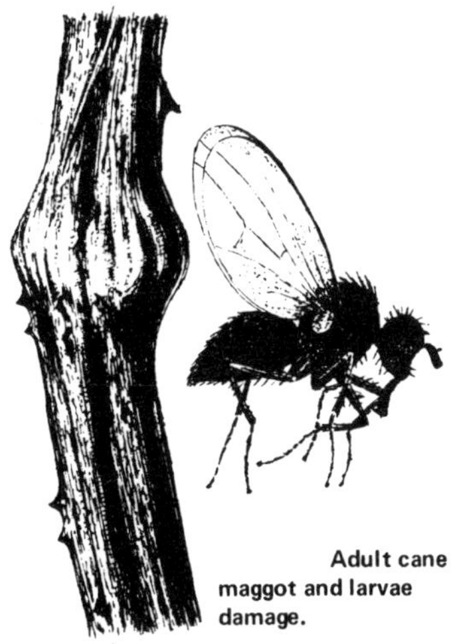

Adult cane maggot and larvae damage.

RED-NECKED CANE BORER

The adult is a blue-black beetle about one quarter of an inch long. The neck of the beetle really is red. It appears on the canes around May or June, lays its eggs on the cane bark, and then the larva tunnel beneath the bark. The boring causes the bark to swell. The swellings contain the white grubs up to one half an inch long. Prune out all canes with abnormal swellings and destroy them.

RASPBERRY CANE BORER

The female is a slender black and yellow striped beetle about one half an inch long. This insect's outstanding trademark is the long antennae. When the beetle lays eggs, it punctures the tip of the raspberry causing the stems to wilt. The grubs bore down into the stem and over-winter on the plant. To control, prune out the infested wilted cane five or six inches below the damaged area. Remove the damaged cane and destroy them.

RASPBERRY CROWN BORER

The foliage turns prematurely red and the canes wilt. The borer is about one inch long with a white body and brown head. The adult looks like a yellow jacket wasp but is a moth and usually makes its appearance in the late summer or early fall. It lays eggs on the foliage and the grubs bore into the crown. The feeding habits are common to all borers. It can be found in the crown sections, the larger roots, and the base of the canes. They winter under the bark just below the surface of the soil. In the spring the larvae move upwards attacking the new canes. It takes about two seasons to completely develop. To control, remove infested areas and destroy the canes. You can use a soil insecticide at the base of the plant.

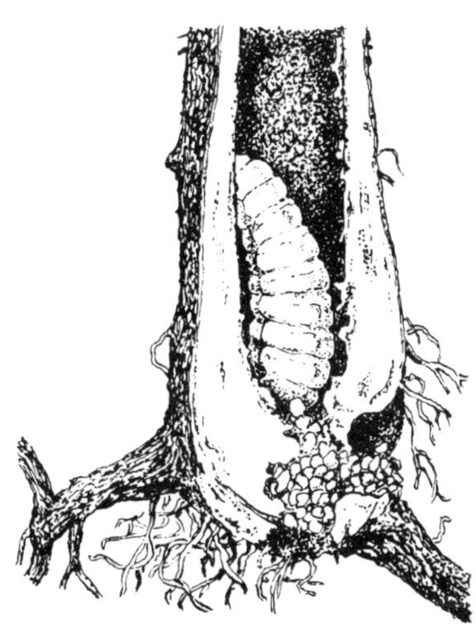

Raspberry Crown Borer and damage to the crown.

RASPBERRY SAWFLY

A foliage eater and common pest. The worms are spiny and light in color. The adult is a black wasp with yellow and red markings. The larvae eat along the edges of the leaves and can do a lot of damage in a short time.

MITES

Sometimes during very hot weather, spider mites can be seen on bramble fruit. The foliage becomes spotted brown or white. If the damage is severe, the leaves will begin to fall and the fruit will dry up. The mites feed on the underside of the leaves and can be controlled by an organic dust before the fruit sets.

RASPBERRY FRUIT WORM

This grub is whitish and about one quarter of an inch long. It feeds inside the buds or on the developing fruit. The fruit will drop off the vines if the infestation is heavy. The beetle (see drawing) is very small, brown, and hairy. The beetles are heavy feeders on the new green growth and will skeletonize a plant if not controlled. They lay their eggs on the buds near the new fruit. The mature grubs drop off the canes, and over-winter in the soil, and emerge the following season as beetles. To control the raspberry fruit worm, cultivate the soil around the brambles in the late summer to destroy the pupal cases.

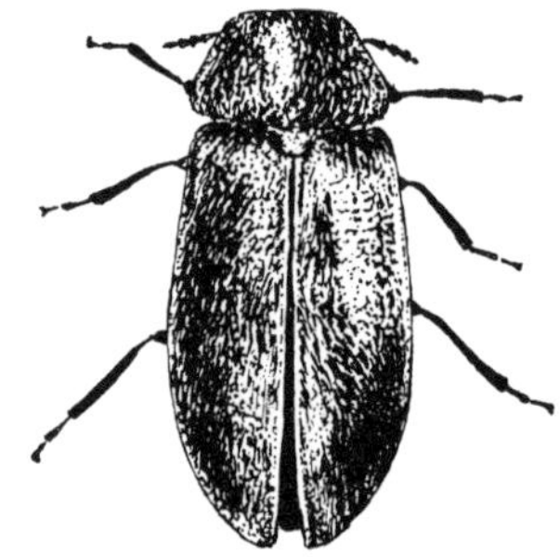

Adult Raspberry Fruit Worm.

PICNIC BEETLES

Sometimes called sap beetle, they are black with red and yellow spots on their backs. They like over-ripe fruit, so become a nuisance during the harvesting period. For best results in controlling the picnic beetle, remove all damaged canes in the region of the diseased fruit. Pick the fruit as soon as it becomes ripe to avoid most of the trouble.

CONTROL

. Plant clean stock from a nursery or known mail-order house.

. Keep weeds down in the patch and remove any wild brambles in the garden site.

. Remove all canes that have fruited and prune out weak canes.

. Keep patch thin. Good air circulation is important, all diseases like a moist atmosphere.

. If you use chemicals (fungicides for diseases) develop a regular schedule for your area. I suggest you ask your local agricultural extension service for the proper chemical for your microclimate.

. For good cultivation it is important to understand the cycles of insects and diseases, and to know when to prune and remove canes (before or just as the disease strikes). Healthy plants resist disease, so follow a good sound fertilizing program. You can get a soil test for your region and determine the nutrient levels of the needed food for the plant.

ANTHRACNOSE (See drawing)

This fungus attacks all raspberries. Small purple spots appear on the canes causing them to dry out and crack. The fruit will not develop properly and may dry up. Black raspberries seem to be more susceptible than the red types, so plant black raspberries in a different location than the red.

CROWN GALL (See drawing)

Swelling on the root sections which are well over one inch. This is a bacterial disease.

Raspberries grow best in full sunlight with plenty of air ventilation. Separate the new patches from any wild berries and plant in sites not previously cultivated with berries. The fruit will open faster on a southern slope

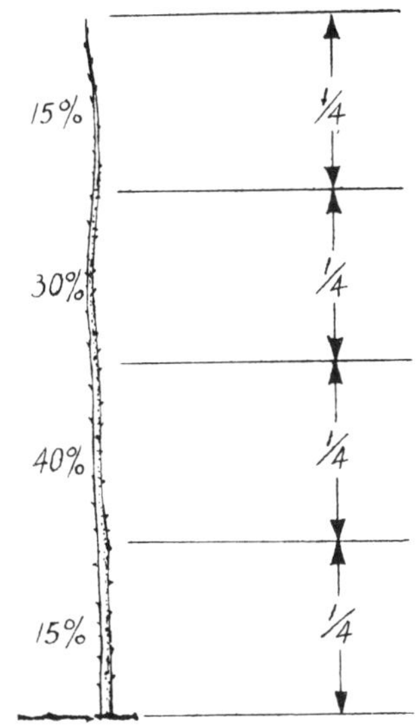

"Heading back" a red raspberry cane. Here you see the average percentage of total production by each quarter of the cane. Never cut off more than the top quarter.

Anthracnose

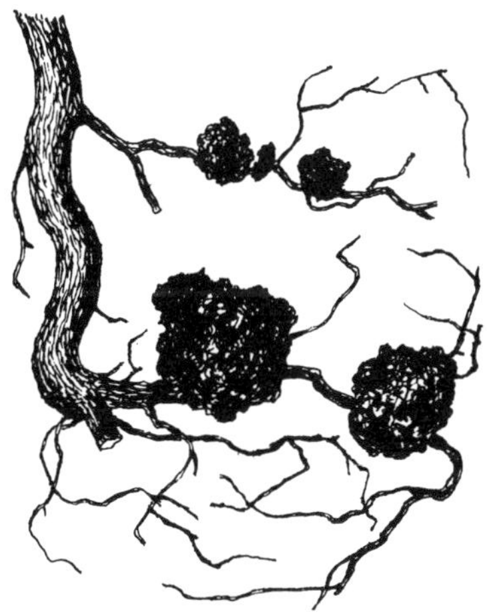

Crown Gall

(which is not as wet) thus preventing some fungus attacks. Try and grow these varieties: June, Latham, Fall Red, Viking, Taylor, Newburgh, Hilton, Hertitage, Blackhawk, Cumberland, Sodus, Marion, or Clyde.

LATE RASPBERRY RUST

In late July or August this fungus may cause premature defoliation. Small yellow blisters appear on the underside of the leaves and on the canes.

LEAF CURL

Another virus carried by aphids. The symptoms are a downward curling of the foliage and shortened internodes.

MOSAIC (See drawing)

This probably is the most common of all the viruses that attack the bramble fruit. Like most viruses it is carried by insects (primarily aphids). Maintain a good insect control program and keep weed plants out of the patch. Buy "virus free" plants from any established nursery.

ORANGE RUST

Attacks only purple and black raspberries. This appears as blisters on the underside of the leaves in the late spring and early summer. The canes that bear the infested foliage look weak. This fungus will spread throughout the plant, including the roots.

POWDERY MILDEW

A grayish-white covering develops on the foliage during humid wet weather. Grow the brambles out in the open, in the hottest part of the garden. The red Latham variety is very susceptible.

SPUR-BLIGHT (See drawing)

A serious fungus disease of red raspberries. Look at the leaf and cane where they meet and if you notice small purple spots, you probably have leafblight. These spots enlarge and may become two or more inches long. When the infested fruiting canes begin their growth next season, the foliage is pale and yellow and the canes will split. The fruit is small and hard. Follow pruning and good cultural progr and avoid most of this damage.

STREAK

A purple streaking of the canes. The infested canes will curve down by the end of the summer. Streak only bother black raspberries.

VERTICILLIUM WILT

The symptoms of this fungus differ with locale, weather conditions, and the variety of plant you are growing. Symptoms are:

1. Dull green leaves on the lower cane instead of the normal green healt plants.

2. As the disease progresses the leav turn yellow and drop off.

3. The top of the cane wilts and the entire cane will die.

4. "Blue stem": This is a dark blue that may cover the entire cane or appe as streaks. This symptom appears on the roots or near the soil line. It is a severe vascular disorder and the stems look water-soaked.

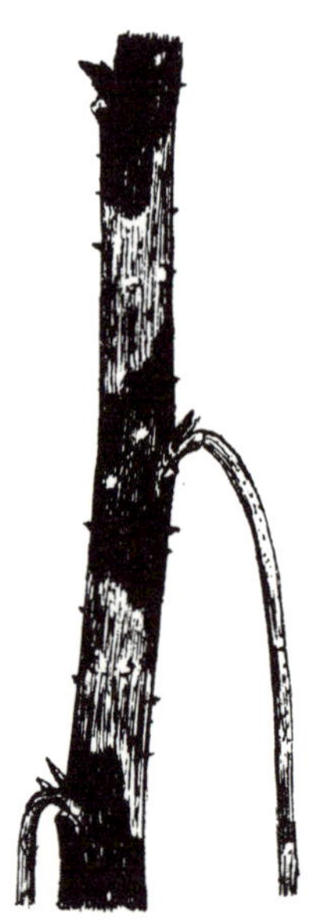

Spur Blight

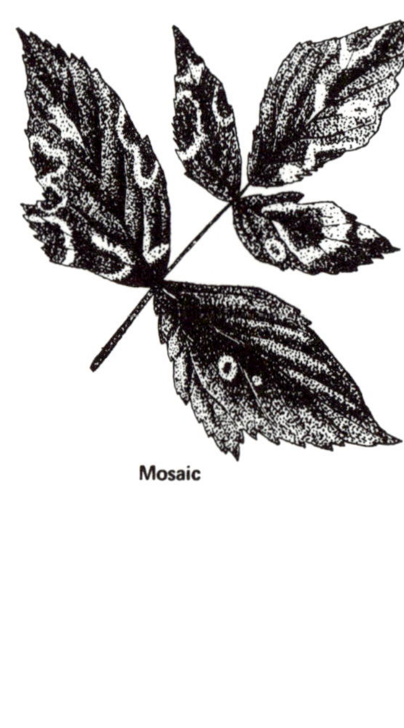

Mosaic

5. Buds that look healthy will open but the new growth will turn yellow and die.

6. When and if berries develop, they will remain small and poor in quality.

7. Blackberries and raspberries both have about the same symptoms. This soil borne-fungus enters the plant through the roots. Buy certified plants from a nursery or commercial propagator. This is a very common wilt, and infects vegetables (eggplant, tomatoes, and potatoes), strawberries, and all brambles. Try not to plant in the same location in the garden if you buy or propagate new canes every year.

YELLOW LEAF RUST

Appears mainly on red varieties. Yellow blisters appear on the leaves and they drop off. In the fall the yellow will turn to brown or black. Orange rust and yellow leaf rust are very similar. Remove spotted canes and destroy infested parts of the plant.

PESTS OF GRAPES

Anyone who has propagated grapes and worked in vineyards knows the number of pests that a single grape vine can attract. I will not attempt to list all the insects that attack the grape (there are books on Viticulture for those really interested in exploring the grape), just those I think the most common and that seem to do the most damage.

GALL MAKERS

A variety of insects and mites make galls on grape plants and live inside these structures. The larvae are usually red, orange, green, or yellow. They are tiny and cause unsightly swelling of foliage and stems. These insects are subject to attack from natural enemies. Prune out the infested canes and remove all leaves with galls on them. If it gets to be a very serious problem consult a professional horticulturist in your area. Many grape varieties are resistant to some gall-making insects and diseases. Your fruit catalogues or the tags on the vines will list them.

GRAPE BERRY MOTH

The adult is a brown moth with a half inch wingspan. It appears late in the spring or early summer. The larva is a green caterpillar about three eighths of an inch long. It builds a cocoon late in the summer which falls to the ground in the fall. The larva feeds on the flowers, new fruit, and later in the season on the pulp and seeds. Found in the Eastern United States.

You can control this pest without the spraying of chemicals. The first method involves clearing the soil from the base of the plant about a month before harvest. Form a slight ridge at the base of the plant. The cocoons of the grape berry moth will fall on the mounded up soil in the late fall. In the spring, pull the soil away from the base of the plant and level the soil, to separate the cocoons from the soil. You will be able to see and discard the pest. This operation should be done before the plant sets fruit blossoms. You can also cover the area under the trellis or fence with a black plastic. Keep the ground level, or slightly lower than usual, and in the early spring, cover the fallen insects. The soil must be sealed compactly or the moth will emerge. The basic principles employed in both systems are cultivation and observation. If you understand the habits of the moth and the various cycles, you will be able to work out a control system that suits your needs.

GRAPE FLEA BEETLE

The adult is a shiny dark-green jumping beetle about one quarter of an inch long, which eats the new buds and the new foliage. It is found in the East and parts of the Mid-West. Hand picking seems to be the best method of killing this beetle, if you have a few vines and not a wine-making operation. The grape flea beetle likes sheltered areas like wood lots or brush-cover. If you plant the vines out in the open (grapes need a lot of sunlight anyway), and away from wooded areas, you will avoid most of the beetle population. They hibernate under fallen leaves and near trash, so keep the garden clean.

GRAPE MEALYBUG

This form of the mealybug is white with a cottony appearance. It grows to just over a quarter of an inch long. Like all sucking insects, the mealybug secretes a honeydew that attracts ants and on the waste, sotty mold develops. Found in all parts of the country. All the sucking insects are prone to attack by the lady beetle. If you do not wish to buy lady bugs and let them loose in your garden (and for the life of me I can't see why not, unless you just do not like the idea of insects), you can simply wash down the plants with a good heavy spray of water.

GRAPE PHYLLOXERA

Looks like an (see drawing) orange to a reddish tan aphid with wings. Is just over one sixteenth of an inch long. The young insect is an oval-shaped green or brown pest which lives inside galls (swelling) on the plant. It is found throughout the United States and Canada. You can grow many disease resistant varieties. This is a European problem and you can avoid phylloxera by grafting resistant American stocks on the variety you wish to grow. Many of the commercial grape propagators do this, so in most cases you are safe if you buy from an established nursery or mail-order house.

GRAPE ROOTWORM

The adult is a big brown, hairy beetle which eats holes in the leaves. The larva is white with a brown head. This insect can kill the plant if the larva infestation is heavy in the soil. It eats roots and weakens the plant. You can control by cultivating around the base of the plant, being careful not to disturb the root sections. You can hand pick the beetles or place a piece of plastic around the plant on the ground and shake the vines. The beetles will drop off-and the rest is up to you.

DISEASES OF THE GRAPE

All of these diseases are covered in other sections.

BACTERIAL CROWN GALL

BLACK ROT

CANKER

FRUIT AND LEAF SPOT

FRUIT ROT

LEAF BLOTCH

MILDEW

NEMATODES

RUST

SHOOT BLIGHT

VIRUSES

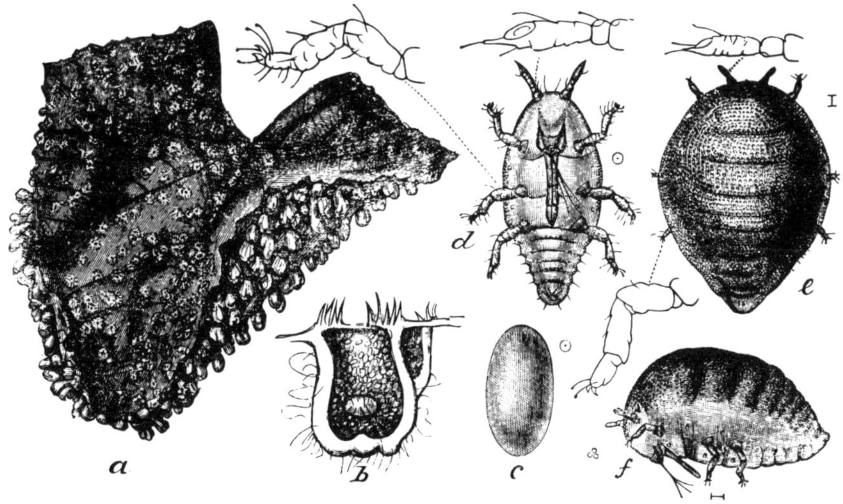

-*Phylloxera vastatrix.* *a*, leaf with galls; *b*, section of gall showing mother louse at center with young clustered about; *c*, egg; *d*, larva; *e*, adult female; *f*, same from side—*a* natural size, rest much enlarged (original).

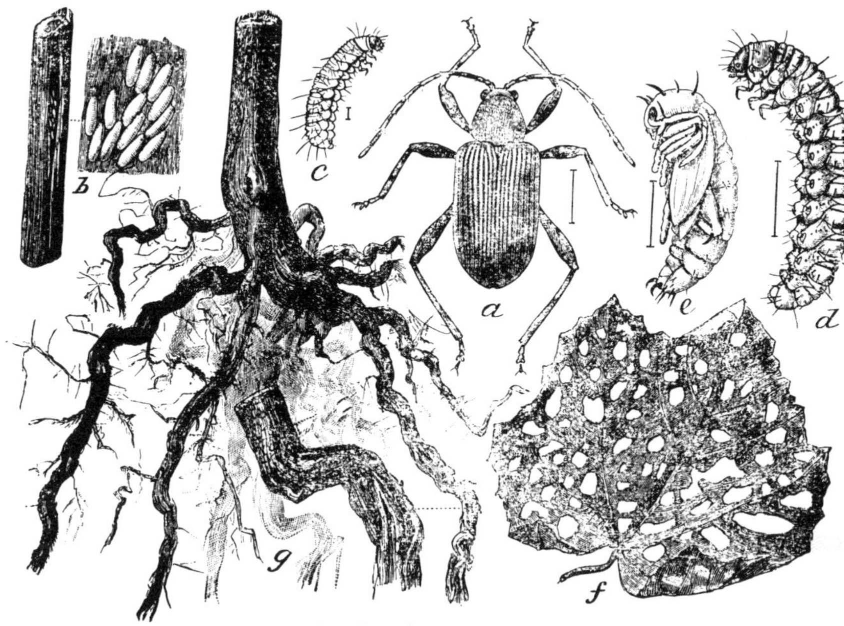

-*Fidia viticida.* *a*, beetle; *b*, eggs represented natural size under fold of bark and much enlarged at side; *c*, young larva; *d*, full-grown larva; *e*, pupa; *f*, injuryto leaf by beetles; *g*, injury to roots by larvæ—*b* (in part) and *f* and *g* natural size, rest much enlarged (original).

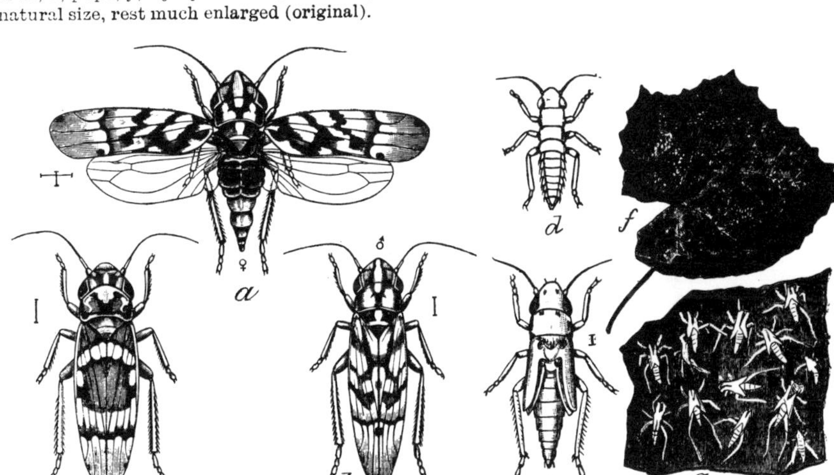

-*Typhlocyba* spp. *a*, *T. comes* Say, female; *b*, *T. comes* Say, male; *c*, typical form of *T. vitifex*; *d*, larva; *e*, pupa; *f*, appearance of injured leaf; *g*, cast pupal skins (original).

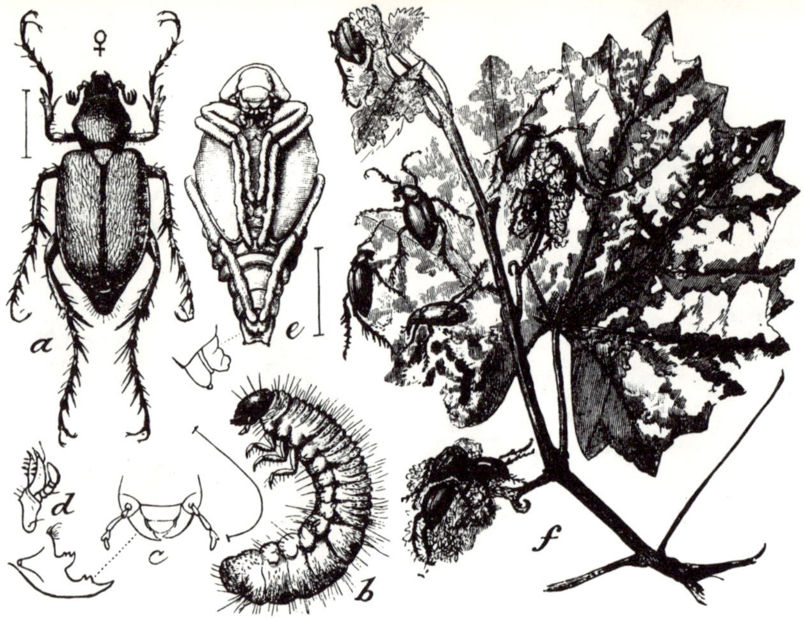

–*Macrodactylus subspinosus.* *a*, beetle; *b*, larva; *c* and *d*, mouth-parts of same; *e*, pupa; *f*, injury to leaves and blossoms with beetles, natural size, at work (original).

INSECT PESTS OF THE GRAPE

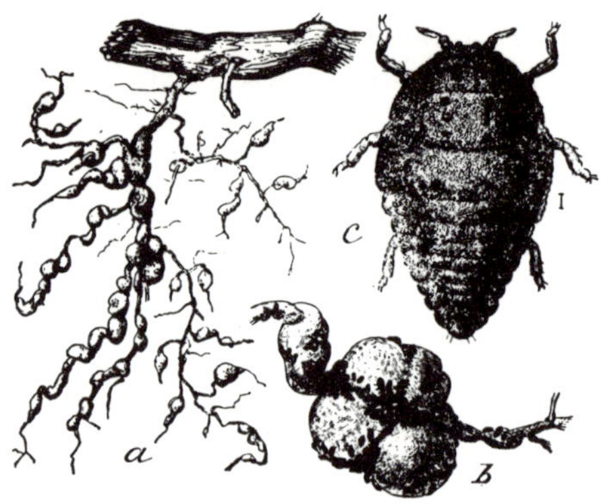

-*Phylloxera vastatrix.* *a*, root galls; *b*, enlargement of same showing disposition of lice; *c*, root-gall louse—much enlarged (original).

–*Amphicerus bicaudatus.* *a*, beetle, dorsal and lateral view; *b*, pupa from beneath; *c*, larva from side, with enlargements of the thoracic feet; *d*, burrow in apple twig made by adult; *e*, larval gallery in tamarisk, with pupa in cell at end; *f*, injury to young shoot and cane, showing the entrance to burrow of beetle near *f* and the characteristic wilting of the new growth—all much enlarged except *d*, *e*, and *f* (original).

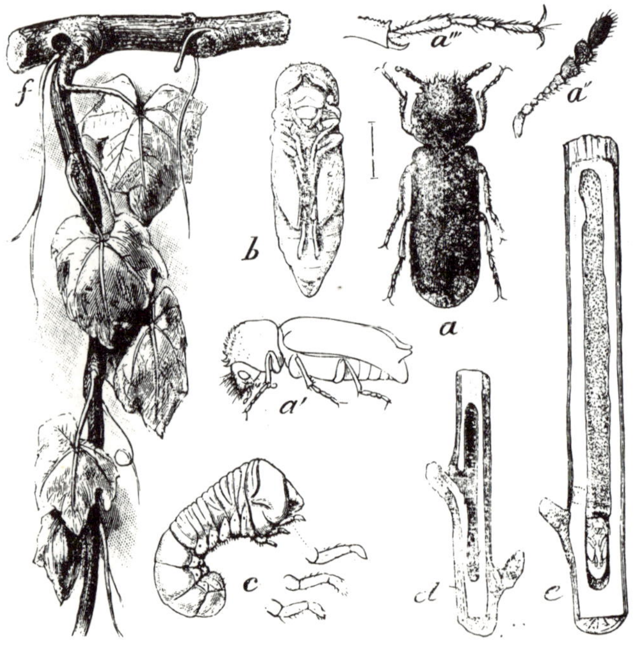

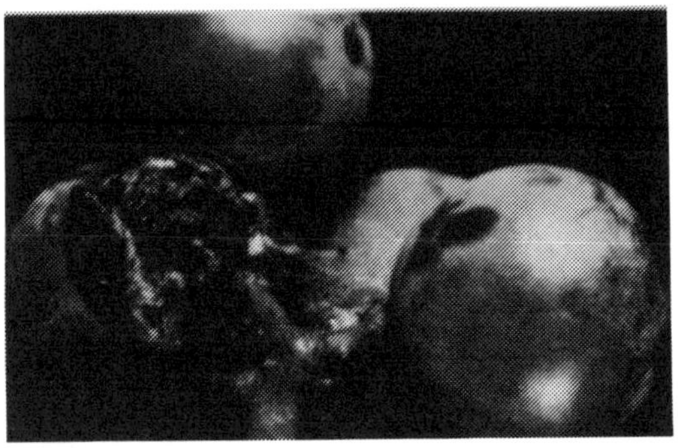

Injury by grape berry moth.

Grape flea beetle larva.

NIAGARA.

Major Insects and Diseases of Fruit Trees
(X = major pest)

Pest	Apple	Pear	Peach and nectarine	Apricot	Plum	Sour and sweet cherry
Insects						
San Jose scales	X	X	X	X	X	X
Apple maggot	X				X	
Borers			X	X	X	X
Codling moth	X					
Mites	X		X		X	
Oriental fruit moth			X	X	X	
Plum curculio	X		X	X	X	X
Diseases						
Apple scab	X					
Blotch	X					
Brown rot			X	X	X	X
Cedar-apple rust	X					
Fireblight	X	X				
Powdery mildew	X					
Peach leaf curl			X			
Peach scab			X		X	
Cherry leaf spot						X

STRAWBERRY

WHITE-FRINGED BEETLE

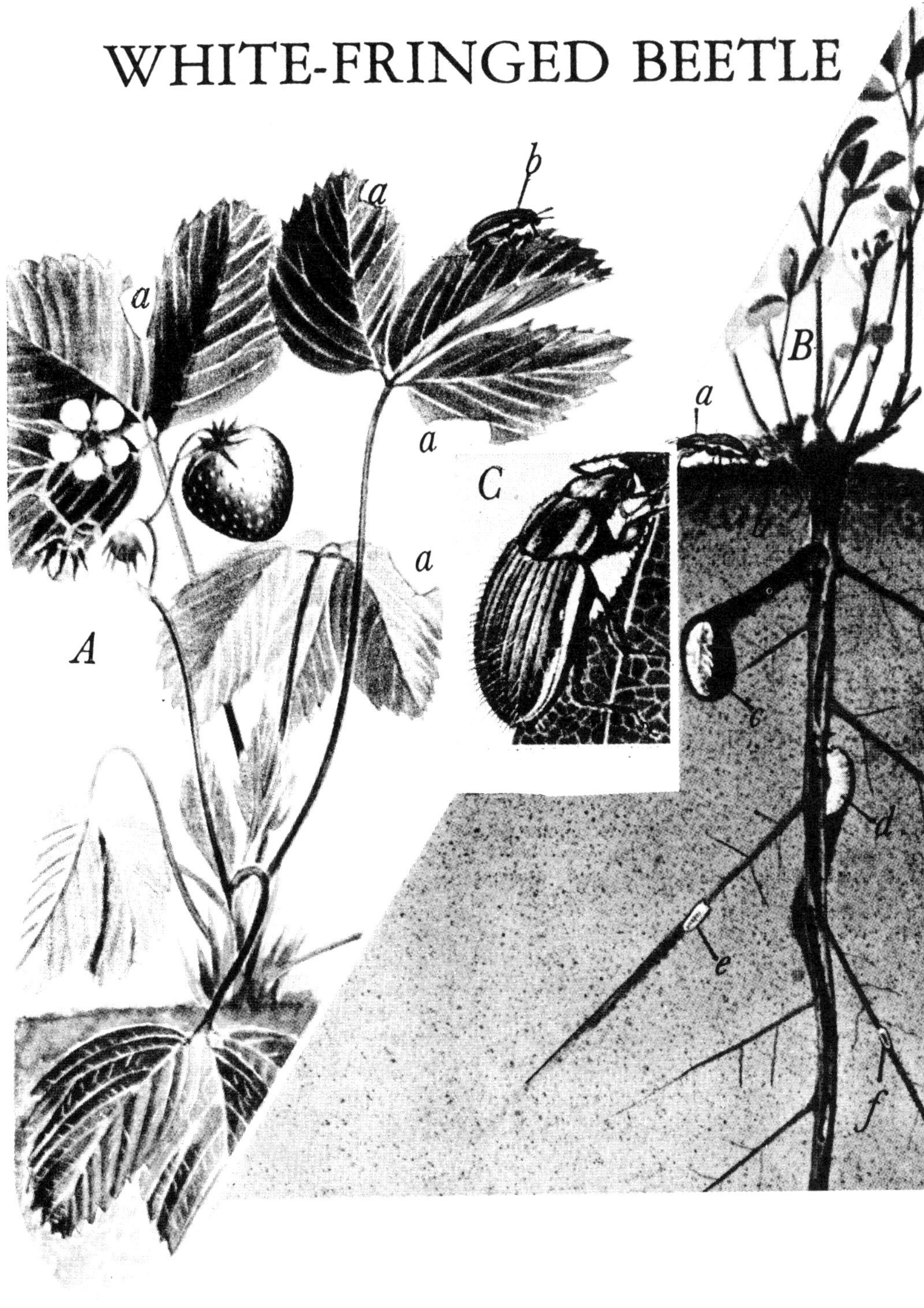

A.a. Strawberry plant injured by white fringed beetles; b. adult beetle; B. Larval injury to alfalfa roots; a. female ovipositing underground litter; b. egg mass; c. pupa; d. full-grown larva; e. and f. immature larvae; (all enlarged).

Adult

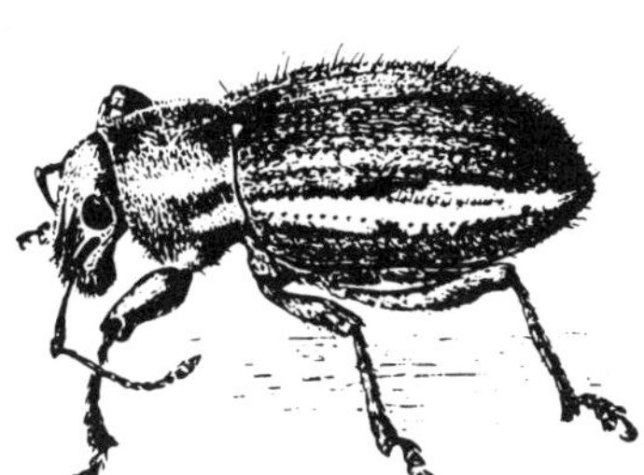

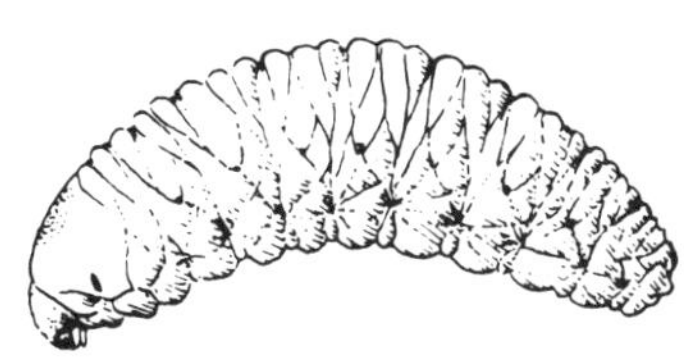

Larva

SHADE TREES, MINOR SHRUBS, AND PALM

REPAIRING TREE INJURIES

If you make pruning a regular part of your yearly landscape maintenance, you can improve the health and appearance of your garden and prevent a lot of unnecessary problems. This is true for shade trees as well as small ornamental trees such as flowering crabs, flowering cherry, dogwood and all of the standard and dwarf fruit trees in your garden. If you prune when the problem becomes apparent, it will eliminate major tree surgery later. For example, emergency pruning sometimes is necessary just after an ice storm or wind storm.

Pruning trees is a part of preventative garden maintenance. If a plant is kept correctly pruned, you can control the growth and not have the garden take over. Once a plant has over-grown it is sometimes difficult or even impossible to see the insect or disease outbreaks. Plants not only need to be pruned for health, they also look better correctly trimmed. Many gardeners are afraid to cut plants, but professionals will tell you to chop away. Buy a good sharp pair of shears and a good tree saw.

You can prune deciduous plants any time during the year. When the plant is completely defoliated it is easier to see the structure of the tree but you can prune in the summer if the early cold or late frost bothers your tender bones.

Many times gardeners tell me they think a plant is injured if it bleeds after pruning. Bleeding or running of the sap is normal and will not harm the health of the tree. Some heavy bleeders are: dogwood, birch, all the elms, yellowood, and the maples. Healthy trees usually stop bleeding after the new growth begins in the spring. If trees continue to bleed after a normal time period, say after the new foliage is set, they may be infected with a disease called slime flux. Most evergreens can be pruned in the late winter or early spring before the new growth begins. This is true for

TYPES OF PRUNING

BEST WOOD TO PRUNE FOR RAPID REGROWTH

- NEW WOOD
- LIGHT COLOR
- SMOOTH BARK
- NEW BUDS

PROPER PRUNING ANGLE

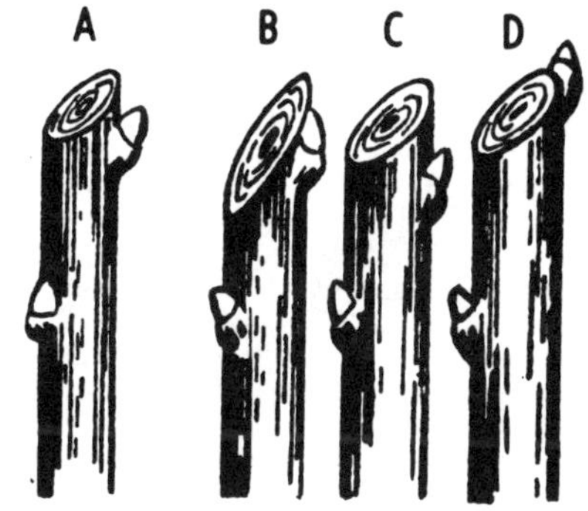

When possible, cut back to a side bud and make the cut at a slant. A is cut correctly. B is too slanting. C is too far from the bud. D is too close to the bud.

You can reduce open spaces on spruces by cutting off one-half of the leader, or terminal shoot, in the spring when the new needles are about half developed.

Keep side branches from growing out of bounds by removing the terminal bud. This not only slows outward growth but also helps to make the plants more bushy.

If the tree develops two leaders, remove the less desirable one in early spring. Trees with more than one leader are weaker and less attractive than trees that have a single, strong, central leader.

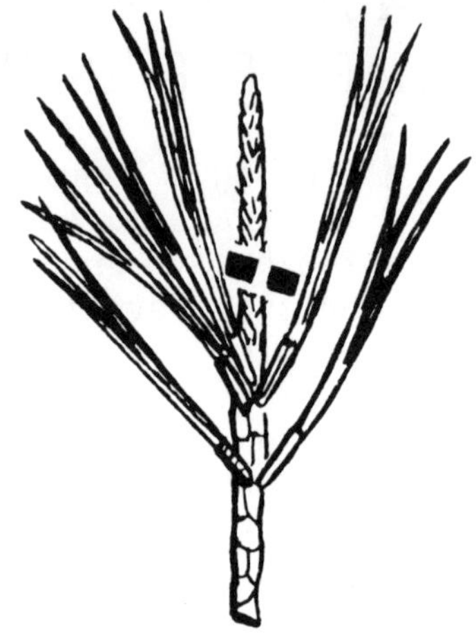

Prune pines in late spring by removing one-half of the candle, or new shoot. Do not damage needle tips because the tips of cut needles tend to turn brown.

You can replace a lost leader by tying one of the branches in the top whorl to a vertical brace.

SHAPING CONIFERS

the long needled (pine), the slender needled (fir, spruce), and the short needled ones (hemlock). Do not prune in the late summer or early fall.

Evergreens of the arborvitae family can be sheared or pruned at any time. One thing to remember when pruning these flat needled evergreens, if you prune in the late summer or early fall you may induce new growth--and if an early frost hits, the plant may be damaged.

ORNAMENTAL TREES AND SHRUBS

All plants that set flower buds in the late summer and produce flowers the following year should be pruned soon after the flowers bloom. The reason is obvious: If you prune at the time the plant is setting buds, the following year you will have nice green plants but no flowers. I have inspected many broadleaf evergreens that customers of mine complained about (when I was in the garden center madness) and finally realized after careful consideration of all the possibilities of failure to produce flowers, that the weekend gardener cut all the buds off last fall. This along with the thousand other such field trips, is the reason I no longer open the store in the morning to answer questions and inspect broken twigs. Examples of plants that produce buds one year and flower the next are: dogwood, all the flowering crabapples, all the flowering cherries, spring flowering magnolias, rhododendron, and the golden raintree.

If the plant produces flowers on this year's growth, you can prune in the late fall or very early spring. Pruning plants every year at the right time encourages new growth and helps establish healthy plants. Many gardeners are afraid to prune, but this is just the opposite of proper care. Take an inventory of all the perennial plants in your landscape and list the ones that require pruning, and try to keep a program of year-round maintenance.

HEDGE PRUNING

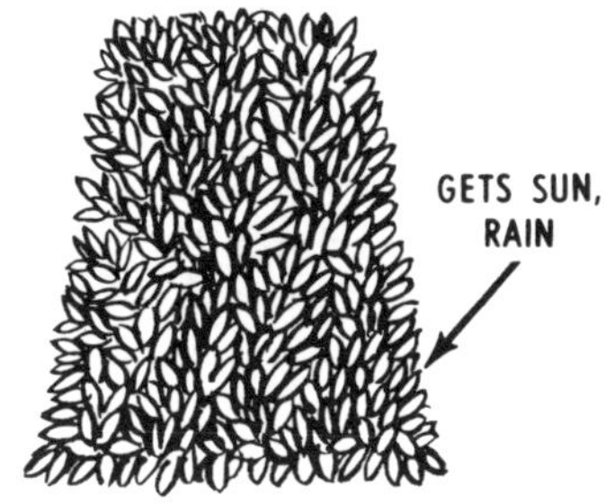

RIGHT

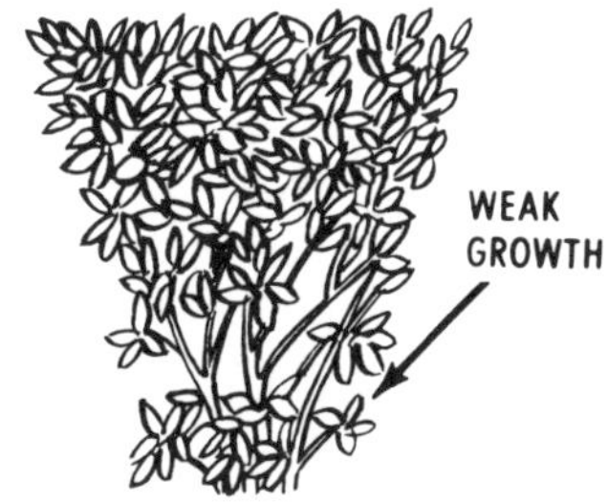

WRONG

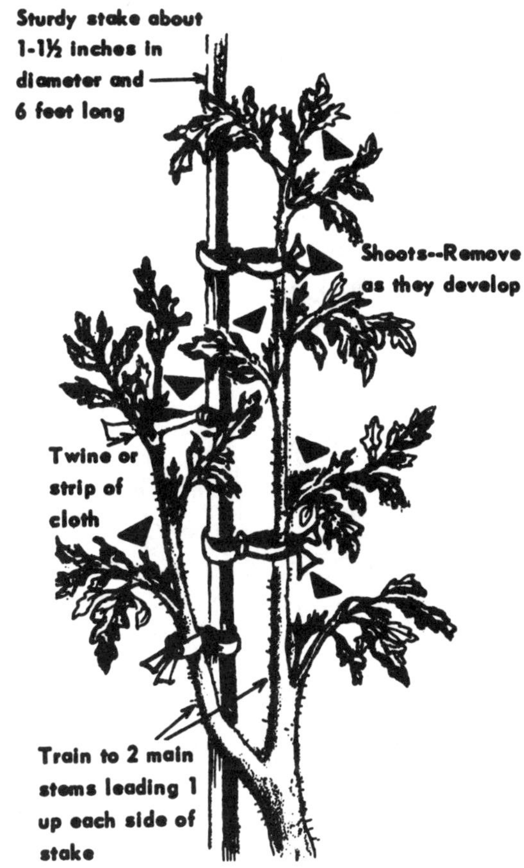

How to prune and tie plants in the home garden

CARE OF PALMS

A short paragraph on the care of the palm tree
general garden books deal with this tree, but many Americans now live in regions where the palm is as common as the maple in the Northeast. Here are a few tips that might make your palm tree healthier.

1. Never prune out the terminal bud on a palm tree If this happens the whole stem will die.

2. Multiple-stemmed palms produce new growth from below the surface of the soil. Prune out the older stems and the plant will not only look better but will react in a more healthy way. Cut these old stems off very close to the ground. It will not damage the plant.

3. Remove all diseased parts of the palm.

4. Dead leaves of most palms form a shirt around the trunk beneath the crown. These groupings of dead leaves are not only a fire hazard to the home owner but a gathering place for insects and it harbors diseases. As soon as leaves begin to droop remove them. If they become discolored remove them.

5. Many palms, like the royal palm, she their leaves. If they fall they may injure someone, so I would suggest removing them before they get too bad.

6. When removing palm foliage always cu the underside to prevent tearing the fiber of the plant.

CARE AND PRUNING OF OLD TREES

Many times when people move into a new home or just want to clear a part of the woodlot, they prune the crowns of the older trees and thin out trees around the plants they want to keep. This may result in a serious condition called sun scald. After the bark is suddenly fully exposed to bright direct sunlight, the bark is often damaged. Trees with thin bark are very susceptible to this sudden transition from an indirect sunlight to a bright sunlight. Many of the young maples are very susceptible: sugar maple, red maple, linden, apple, and beech. Some young oak trees may fall into this category in their early stages. To prevent this trouble, thin out the crowns of the old trees over a period of time. Never do it all at once in the spring or early summer. If you are thinning out a lot, take only a few trees out each year and never cut all at once. Not only will you damage the trees that are left, but you may cause soil erosion in the immediate area.

Pollarding is another form of pruning. This is the cutting of the tops or crowns of large trees. This should be done only in unusual cases, say for example, if telephone or other electrical wiring needs to be hooked up. Never cut the tops of trees except in an extreme case. Some trees grow back in a hurry (like willow or silver maple) but usually the plant is ruined and will never regain its original shape or health. You should seal the wounds or pruning cuts to protect the tree from insect infestation and diseases.

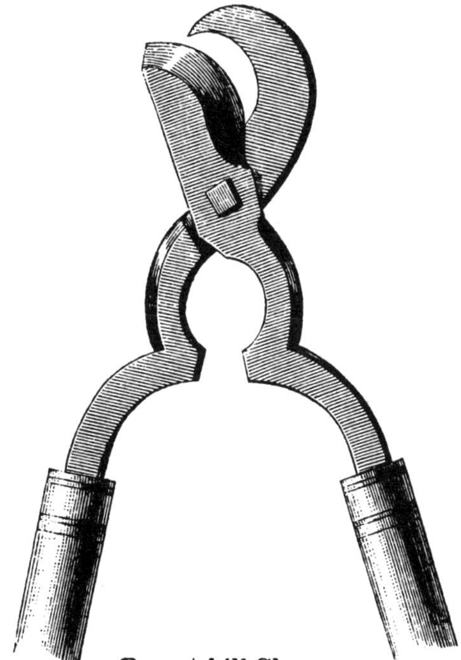

Parrot-bill Shears

"Trench - In" long-stemmed plants

PRUNING SHEARS

Save Your Trees

PRUNING SAW

LOPPING SHEARS

SHADE TREE DAMAGE

BARK

When a serious injury occurs on the bark of a tree (like a truck hitting it or a major car scape) you can follow two procedures. The first, if you know the **injury just occurred, is to hold the** bark in place. You can do this by nailing the bark down- as long as the inner cell structure on the bark has not died. Wrap the wound with moist moss and cover with a plastic wrapping to keep the moisture in. Seal the edges of the plastic with an asphalt paint, then bind the whole thing with a heavy cord. In a short time you will be able to see if the healing took place. If not, try this: remove all the bark that is damaged, and apply around the edge of the wound any tree paint that is antiseptic. Try the first and if you don't succeed try the second. If you notice a break in the paint surface, repaint. Many wounds take a few years to heal properly.

These storm-damaged branches should be removed and the wounds smoothed, shaped, and coated with wound dressing.

LIGHTNING DAMAGE

Sometimes it takes a long time to tell if a tree has been damaged seriously by lightning, except for the more obvious damage like a splintered limb or broken branches. Paint exposed wood and remove all damaged limbs. I suggest you call a professional tree man for big jobs. Cutting big trees that have been hit by lightning is very dangerous, and often you cannot see all the trouble. Only an experienced tree worker can spot real trouble. Prune the roses and dogwood and whatever you can handle, but let a professional do the dangerous, heavy work.

Crushed by storm, the tree still shoots upward.

SPLIT TRUNKS AND CROTCHES

The best thing you can do for a small split in the branching section of the tree is try to restore it to its original position and hold it there. If this can be accomplished before the limb becomes brittle and the cell structure dies, it will save the appearance of the tree and a lot of work and maybe money.

SPLIT TRUNKS

Smooth the edges of the damaged parts and disinfect the wound with a denatured alcohol or shellac. Drill holes directly through both parts of the split in a straight line. Then put bolts into the holes and carefully draw the split section together. Cover the cracks with an antiseptic paint.

SPLIT CROTCHES

The same procedure is followed with a split crotch, although it's a little more complicated. Draw the crotch together with bolts. Additional support may be needed from lagthreaded screw rods or from cables with toggle bolts. This is not an easy job and if the job looks hard, at least consult an expert for the proper procedure if you plan to do the work yourself. Good judgement is needed to determine the number of lag rods or cables needed to properly support the limb, and determine the exact height above the damaged area to place the hardware. Tree surgery is a special branch of gardening and not enough is said about good tree surgeons and professional aborists. I would suggest a good textbook on tree surgery if you plan to do a major job. This is a science and a very specialized craft that I cannot do justice to here. When you are trying to save a plant that is over one hundred years old and has been part of the landscape for a long time, get the best advice you can.

Split crotch repaired by placing threaded rods through both members.

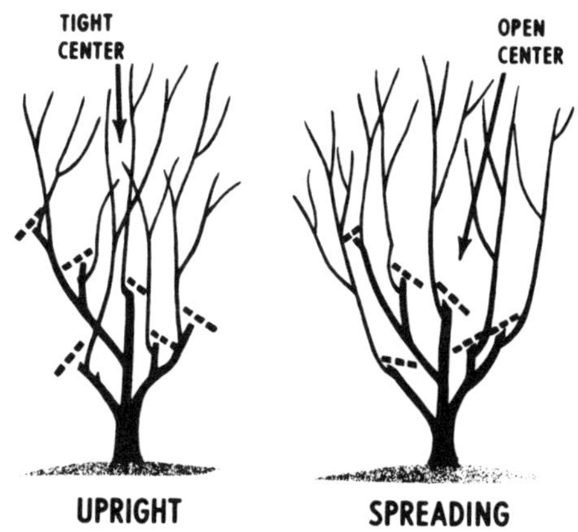

To change the shape of a plant, cut it back to where a branch or twig grows in the direction you want the plant to grow.

You can control the direction of the new growth by cutting back to a side bud that points in the direction you want the branch to grow. New growth will follow the dotted line.

SPECIAL PRUNING PROBLEMS

SHAPING PLANTS

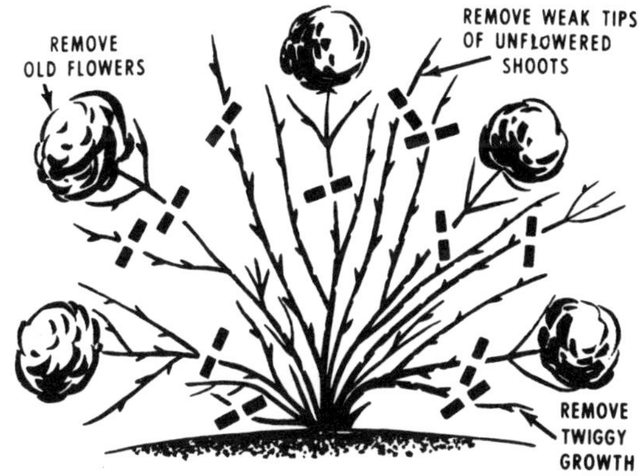

UPROOTED TREES

You can save some trees that have been uprooted if you act promptly. The first thing is to cover the exposed root sections to prevent them from drying out. This is done as soon as you see the exposed roots and before you begin repairing the damage. If this is not done immediately, forget it and just pull the tree out, for if the roots dry up the plant will die. You can cover the roots with soil, wet burlap, peatmoss that has been watered down, or anything that will hold moisture. The procedure is as follows:

1. Cut away broken roots and dress the wounds with an antiseptic asphalt paint.

2. If the tree is too large to pull upright by hand, without doing more damage to the plant and to yourself--use a block and tackle, a dragline, or a bulldozer (if you just happen to have one handy). Pad the trunk of the tree before you move it with machinery or block and tackle.

3. If the uprooting occurs during the growing season and the plant is heavy with leaves, spray the plant with a wilt-proofing material. You can find this product in most garden centers. If not, telephone a commerical nursery and ask to buy some. Even though most growers are not retail outlets, they will always help in the case of an emergency. They use wilt-proof everyday as they dig plant material. All big plants that are dug during the hot seasons are sprayed with wilt-proof as a matter of procedure. This is very important and may just save your tree.

4. Water the plant heavily until new roots develop on the broken root stocks.

5. If the tree looks a little low, fertilize with a special fertilizer for trees or evergreens.

SPRING FLOWERING SHRUBS

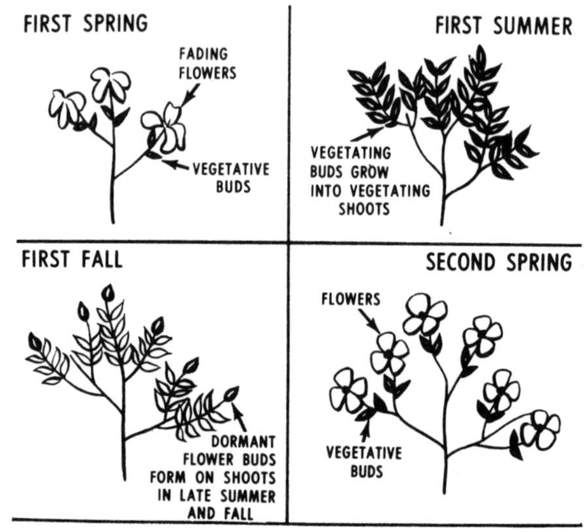

Prune only before vegetative buds begin to grow. Cutting flowers from plants is a good way to prune. Do not prune in fall or winter because this will remove flower buds that would produce flowers the following spring.

REMOVING SEED PODS

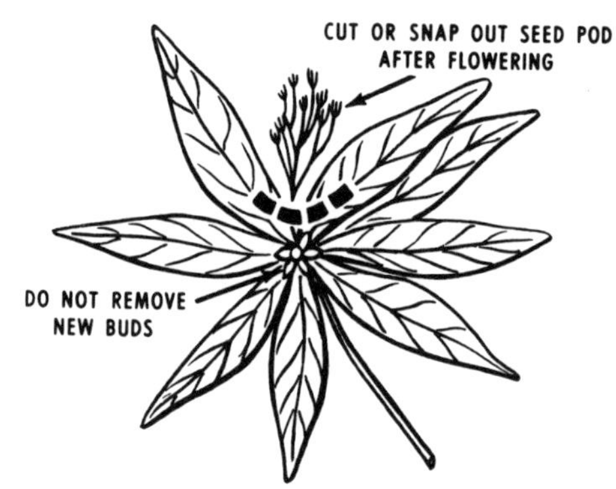

SUMMER FLOWERING SHRUBS

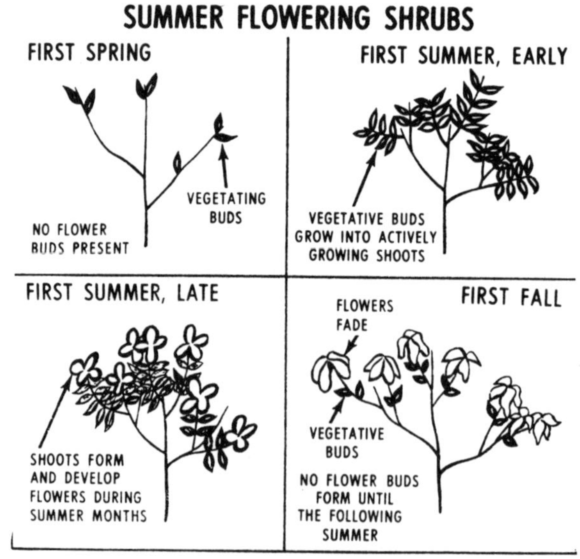

Prune in late fall to early spring before growth starts. Pruning after growth starts removes flower buds that would form blooms.

6. After the tree has been re-set, follow the instructions in the drawing for staking and placing guide wires on the tree. Depending on the size of the tree, always use at least three lines and more if the plant is large. Place the guide lines at least two thirds of the way up the trunk, and anchor the wires in the crotch. Use an old rubber hose cut in sections long enough to go around the trunk of the tree. (see drawing) This is merely a preventative measure to insure the trunk from being cut by the steel wire. Loop the hose-covered wire around the trunk and twist the end of the guide wire back around the main part of the wire. When you put the hose-covered wire around the trunk, keep it a little loose- a little rubbing is better than having the trunk restricted. Hook the wires on the guide stakes in a triangle around the tree. Tie the wires against prevailing winds in your region. Depending on the size of the tree and the damage to the roots, leave the wires and the guide lines on the tree for at least two growing seasons.

PRUNING OVERGROWN PLANTS

RIGHT

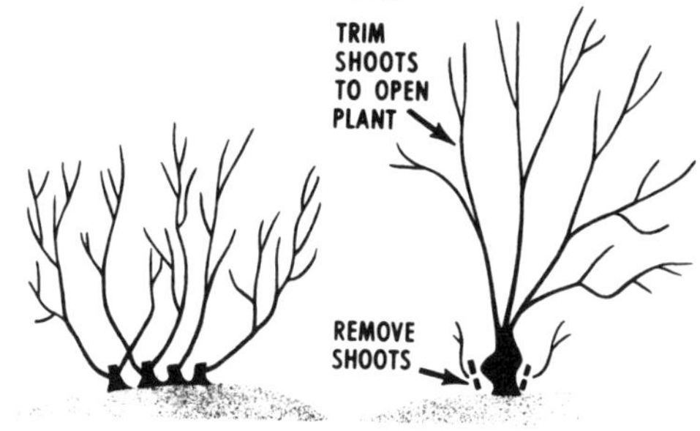

OLD SHRUBS

To prune old shrubs, cut the old stems back to the point where the branches originate, near the ground.

GRAFTED SHRUBS

When pruning shrubs that have been grafted, always remove new twigs that start below the graft knuckle.

WRONG

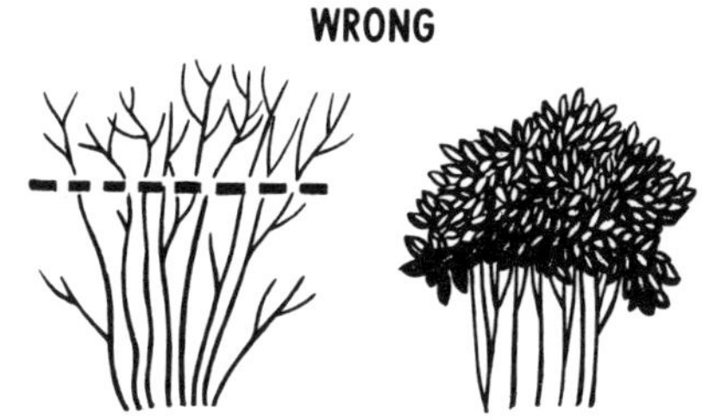

When shrubs are beheaded, as at left, new growth comes only from the top of the plant, resulting in the leggy, bushy-topped shrub shown at the right.

Transplanting Carriage

Dig hole for tree large enough for the roots and plenty of good soil.

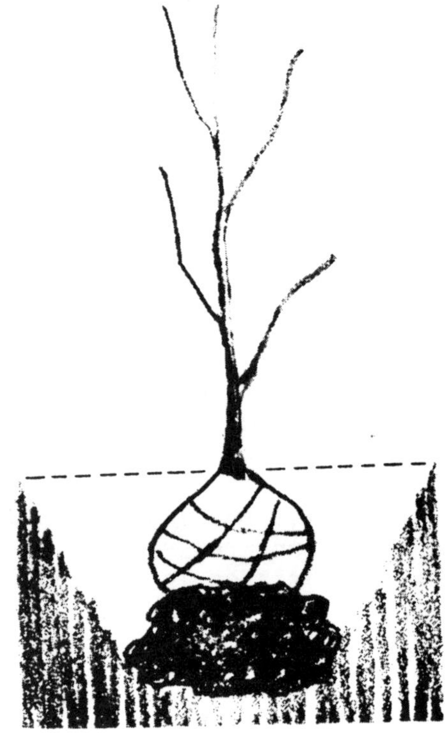

Partly fill hole with good soil so top of root ball will be even with soil surface.

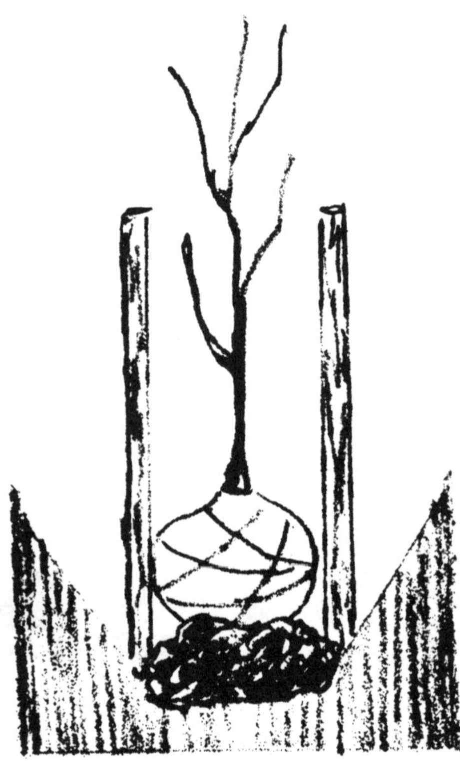

Drive supporting stakes firmly into the soil next to the root ball.

Finish filling hole with good soil, tamping it firmly. Ridge the soil so it forms saucer-like basin.

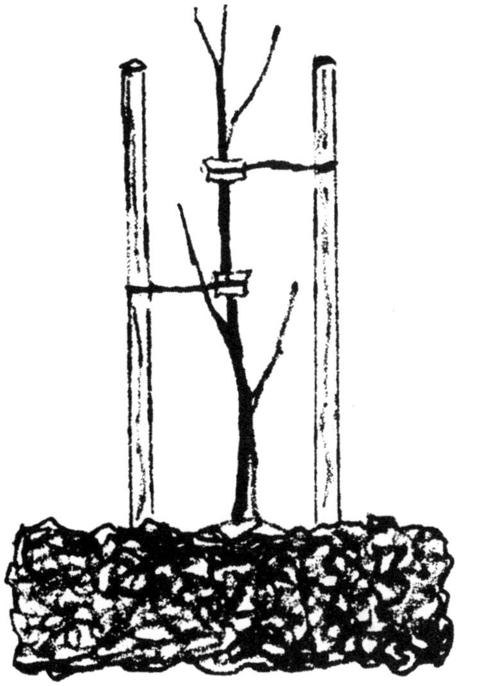

Tie tree to the supporting stakes with hose-covered wire. Water well to settle soil.

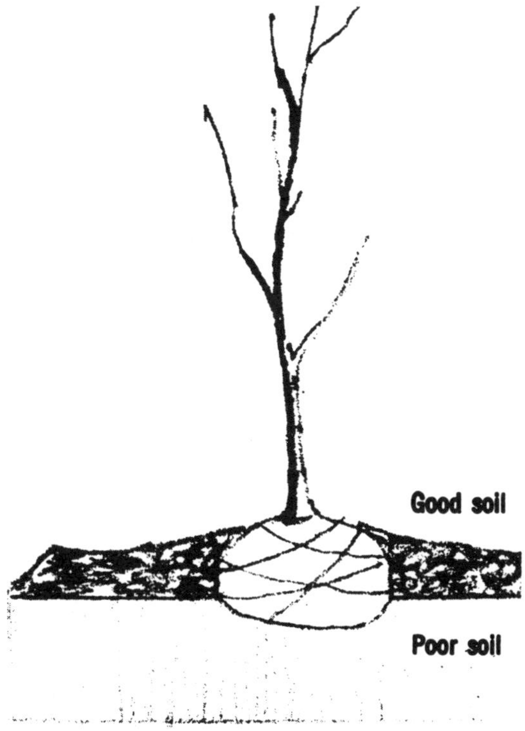

Where drainage is poor, plant trees on the "high side" by building up the area around the root ball with good soil.

For older trees, punch holes for fertilizer with a crowbar—15 inches deep, 18 to 24 inches apart—under the tree drip line. For new trees, scatter fertilizer around the tree at the tree's drip line.

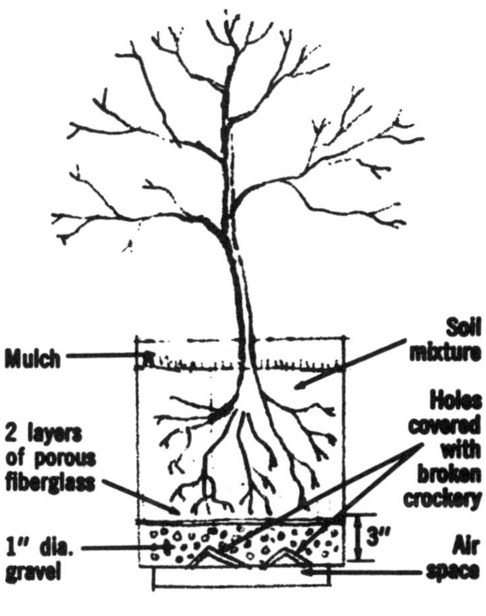

PESTS OF ORNAMENTAL TREES, SHRUBS AND GARDEN FLOWERS

ORNAMENTALS

—A bed specially prepared for plant growth where the normal drainage is inadequate

ORNAMENTALS INCLUDE FLOWERS, FLOWERING TREES AND SHRUBS, SHADE TREES, CONIFERS (BROADLEAF AND NEEDLED), BULBS, AND GARDEN FLOWERS (ANNUALS AND PERENNIALS).

INSECTS

Pests of ornamentals cause several kinds of injury depending on local weather conditions, the feeding and egg-laying habits of the insects, and the location of the plants they attack. Some insects feed on the plants as soon as they emerge from the soil, others feed on the seeds, and the root systems, while others feed on the mature plant.

Injury to the foliage may consist of mining the interiors of the stems and leaves, eating all of the foliage, or just skeletonizing parts of the leaf. Injury to the terminal shoots may be caused by chewing or tunneling into the bud or stem. Some insects hollow out the interior of the terminal buds and the new shoots. A few make pits in the surface of the bark of the main stem after feeding there. Still other insect pests suck the juices from all parts of the plant --stem, foliage, and buds. Insects of the sucking families (like the aphids or scale) leave a striped appearance on the part affected. The soil-living insects (the grubs and certain borers) work on the roots and the very tender seedlings. The last major problem is from the formation of root galls, which may appear on all parts of the plant from the roots to the crown. So, as you can see, for every part of the plant, in most climates, and all the seasons, there is a bug for your bush.

The effect of all kinds of insects on your garden and house plants depends on the intensity of the attack and the condition of the plant. It follows that a swarm of beetles or aphids will do a lot more damage than a few and that if a plant is in poor health, say lacking fertilizer, water, or proper humidity levels (as in the case of house plants), the plant is more susceptible to attack of insects and diseases.

The most important item any home gardener should understand is the specific type of pest, which plants each type bother, and what the pests look like. The enemies of flowers and flowering shrubs are too numerous to list in one book. If you understand the general rule that certain insects either eat, chew, or suck the plant and learn what effects these feeding habits do to the plant itself, you will be able to control all the pests that fall into each category.

Let me put the plant pests into these groupings for easy identification:

1. Leaf-chewing

2. Sucking insects

3. Gall-forming insects

4. Borers (miners and stem borers)

5. Soil and root eating insects

To control all of the above insects, there are sprays, dusts, liquids, and organic methods such as beneficial insects and natural enemies. The section of this book on organic controls describes the kinds of insects you can buy that eat aphids (like the ladybud) or ways to encourage birds that eat a lot of bugs in one season. If you use chemicals, the section on insecticides and other pesticides will give you complete instructions on application, the proper chemical for each type of pest, and the proper mix. I have not included a general spraying program, for each part of the country has different weather conditions and different growing seasons. Certain bugs emerge at different times in various parts of the country, so I would suggest you seek expert advice from your local nurseryman or government extension service. I will tell you how to use them-you must determine when to use them.

LEAF-CHEWING INSECTS

These insects bite into the foliage and eat the tissue. Therefore the control is directly on the foliage with an end result of poisoning the insect through stomach poisons. The major chewers include: bagworm, cutworm, webworm, juniper webworm, tent and other caterpillars, sawflies, the catalpa sphinx, rose chafer, rose curculio, cucumber beetle, blister beetle, Japanese beetle, flea beetle, and imported willow leaf beetle.

BAGWORM

This is a caterpillar that lives in a cocoon bag with bits of leaves attached to it. As the caterpillar grows, the bag increases in size and by midsummer it reaches about two inches. At this time it hooks on to a branch or twig. The wingless female lays eggs and the winter is passed in the egg stage. The eggs hatch in the spring or early summer depending on the location in the country, and then the worms begin to eat the foliage. Bagworms prefer juniper and arborvitae but will nest and feed on deciduous trees and shrubs.

Wild cherry tree showing loss of foliage after eastern tent caterpillar infestation.

CUTWORMS

You very seldom see a cutworm, for they remain in the soil during the daylight hours and only feed at night. The cutworms do a lot of damage and can kill a plant overnight. They cut off plants, usually very small seedlings, near the soil surface. They also climb and feed on the foliage of mature plants and bore into the new flower buds or terminal shoots.

Cutworms are fat caterpillars, usually smooth and grayish-brown, measuring one or two inches long. They hatch from eggs laid by the brown moth in the summer. By late fall they are full grown and bury themselves in the soil for the winter. The variegated cutworm is the most dangerous to plant life.

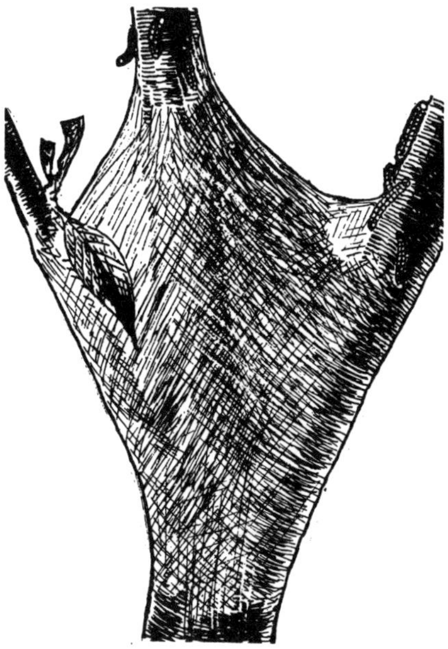

Eastern tent caterpillar nest.

SADDLEBACK CATERPILLAR

Most of our native caterpillars are not poisonous but some do have stiff poisonous hairs that may cause a painful, burning sensation, when you pick up the caterpillar. The saddleback caterpillar, which attacks many kinds of flowers and shrubs, falls into this category. It is brown at both ends with a purple or reddish-blue saddle in the middle.

SAWFLY LARVAE

This larva skeletonizes the foliage of roses and other garden plants. There are three kinds that do the most damage. They are sometimes called false caterpillars or rose slugs. The adults are tiny wasps and the females lay their eggs in slits they make in the leaves. The name is derived from the stiff hairs on the sawfly larva which increases in size over the growing season. Full grown rose slugs or sawflies are about one half an inch long and are yellowish to pale green in color. They have dark green stripes on their backs.

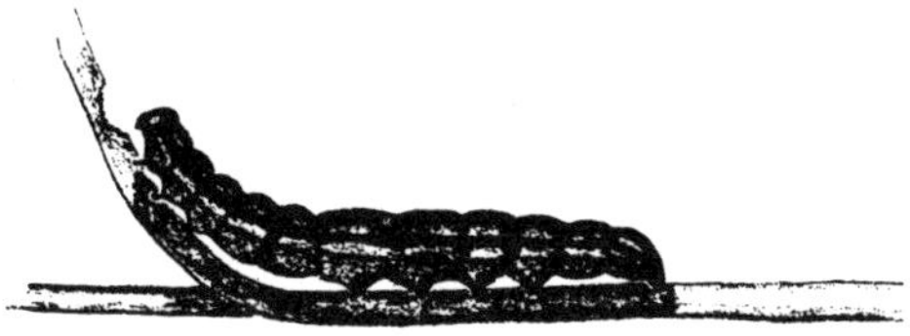

ARMY WORM, *L. unipuncta.*

a. . shows a cluster of larvae feeding on a pine twig; b. adult male; c. and d. are the female and egg niches in a needle respectively. e. mature larva; f. cocoon; (all greatly enlarged).

Close up of a. ovipositing female; b. male; c. female pupa; d. male pupa; e. old egg mass;
Larvae on oak leaf; f. first instar; g. second instar; h. fourth instar; i. sixth instar, or mature larva. Egg masses: k. under branches on rock; l. young larvae spinning down. (Enlarged.)

background showing appearance of tree under attack.
A. entrance holes in bark
B. Inside of bark showing frass-filled tunnels and life stages;
a. egg in niche; b. young larva; c. mature larva; d. pupa; e. adult in nuptial chamber; f. is the entrance hold (all enlarged).

SPOTTED CUCUMBER BEETLE

This insect feeds on the foliage, buds, and flowers, but the major problem is the eating of the holes in the blossom petals. The worst damage is done in the late summer or early fall. At this time the beetles tend to leave the host plant of their desire and migrate to the late blooming flowers like asters and garden mums. The yellowish-green beetle is one fourth of an inch long with 12 spots on the back. The over-winter in the adult (beetle) stage and the eggs are laid in the soil in the spring. The newly hatched grubs (or larvae) feed on the tender new roots before the beetle emerges and begins eating the green parts of your favorite flowers or shrub.

THE ROSE CHAFER

This insect (see illustration) attacks the flowers of roses, peonies, and garden iris. It is a long legged, light brown beetle about one third of an inch long. It usually appears in large numbers in June and July and continues to eat at a very rapid pace for several weeks. If you catch the beetle early in the morning, before they become too active, you can shake tham off the plants and into a pail of water and kerosene. Otherwise you must resort to an insecticide, either spray or dust, chemical or organic.

SUCKING INSECTS

These tiny insects (scale, aphids) get their food by inserting the green parts of the plant into their mouths and sucking the juices into their bodies. This group of insects is usually controlled by contact poison or suffocation. Examples are: aphids, mealybugs, scale, leafhoppers, tree hoppers, psyllids, spiddlebugs, chermids, and spidermites (red spider).

JAPANESE MAGNOLIA.
(MAGNOLIA STELLATA.)

Rose Chafer

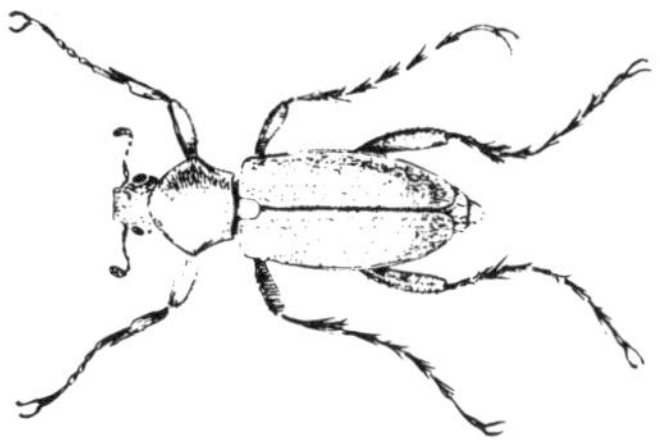

APHIDS (PLANT LICE)

This insect infests every kind of plant except trees. They damage all annuals, all perennials, roses, some flowering shrubs, and indoor plants. A real pest of the first order. They probably will not kill the plant but do cause discoloration of the foliage and the flowering parts, reduce the vigor of the plant by sucking the juices, and may harden the buds. They can be found in large numbers of clusters on the tops of the buds as in roses or on the underside of new leaves. Older and harder foliage do not attract them. Ants feed on the honeydew that is excreded by the adult aphid, so if you see ants on the stems of roses or any garden flower, this is usually a sign that the plant has a colony of aphids feeding somewhere in a nice shady spot on the tips or foliage.

Aphids are soft bodied insects usually green, white, or dark brown. Their bodies are pear-shaped and they have long legs. To see them you must have a swarm for they are very small. Most of the species are without any protective covering but a few species have a wooly covering that looks like wax.

Aphids are disease carriers of mosaic. They attack all parts of the plant- the roots, flowering bulbs, and young seedlings. The green peach aphid is a vector (carrier) of carnation mosaic, lily mottle virus, viola flower breaking, and mosaic diseases of gladiolus.

THRIPS

This sucking insect attacks the flowers, foliage, and bulbs of many ornamental plants both outdoors and underglass. Under the bulb section in this book, I have covered the problems of summer and fall bulbs.

LEAFHOPPERS

This sucking insect causes the leaves to turn a whitish color and curl. It also kills off the tender tops of new plants. The rose leafhopper is a perfect example of damage to tops. The potato leafhopper causes a condition known as hopperburn in dahlias and other garden flowers. The six-spotted leafhopper transmits the virus of aster yellows from plant to plant. This is a very serious condition and you must control the hoppers if you are to control the disease. Copper-containing compounds have been used very successfully against leafhoppers.

Leafhoppers are slender, delicate insects about one-eighth of an inch long. They vary from brown to a pale light green. As their name indicates, they will hop from plant to plant, if disturbed. The eggs are laid in the soft parts of the foliage and many times a couple of generations are hatched each year. The plants and flowers they attack include: asters, calendula, hollyhock, marigold, rose, zinnia gladiolus, and dahlia.

LACE BUGS

This insect attacks perennials like chyrsanthemums, broadleaf evergreens like rhododendron, spring flowering shrubs like azaleas, firethorn, and some types of annuals. There are several species of lace bugs but all have the same patterns and habits. The azalea lace bug is a perfect example of all the other species. It over-winters in the egg stage and hatches in early spring. The young adults (nymphs) are small and shiny. Their bodies are oval and about one eighth of an inch long, and their thin lace wings lie flat on their bodies. Both the adults and the nymphs suck the juices of the host plant. They feed on the underside of the foliage and that causes the azalea leaves to have a spotted or mottled look. The underside of the leaves have black spots that are the excrement of the lace bug. The treatment has to include the spraying of the underside of the foliage on infested plants.

TYPES OF DOUBLE DAHLIAS.

BOXWOOD PSYLLID

This insect causes a cupping of the terminal leaves as a result of their sucking habits. The boxwood psyllid is a small greenish-gray insect related to the other sucking pest, the aphid. The nymphs are covered with a white waxy material and they appear in the spring. If the infestation is severe, the growth of the plant may become stunted, but they seldom kill the plant.

If you apply your controls against the adult psyllid, this may help to break the growing cycle. The adults lay the eggs and if they are killed before that time you will not get bugs the following year. Just as the nymphs start to turn into the adult form is the best time to spray.

PLANT BUGS

This group includes the phlox plant bug, yucca bug, tarnished bug, and the four lined plant bug.

PHLOX BUG

The phlox bug is common and known in most parts of the country. The other plant bugs feed and develop in the same manner, so I will use it as an example. This insect feeds on all the upper parts of the common garden phlox, including the new buds. The injured leaves have a white or light green spotting and the plants become stunted. In severe cases the plant may die.

The adult is about one fourth of an inch long and is easily seen by its orange or reddish wing covers and black stripe on the back. The adults lay their eggs in the summer on the plant, and the winter is passed in the egg stage. In the late spring the eggs hatch, and the nymphs develop to adults in only a few weeks. In some parts of the South you may have two or more generations in one growing season. The best control for long range elimination of this pest is to gather the old plants after the last frost in the fall and destroy them. This will kill the eggs and help prevent the recurrance the following season. Many of the cultural methods that were used before every gardener had a pushbutton spray can really did work. We still have phlox in our gardens after many generations of natural controls.

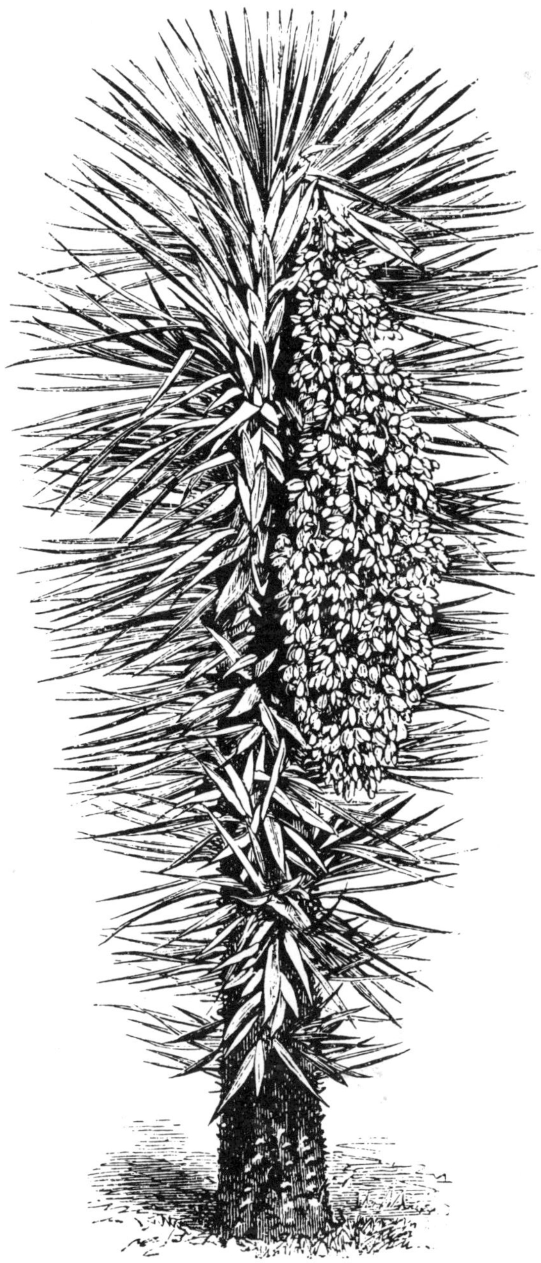

YUCCA FILIFERA

SCALE INSECTS

A sucking insect that falls into two general groups, the soft scale and the hard scale. The soft scale has a rubbery appearance and the hard scale looks like a shield. Scale are very small, usually under one-fourth of an inch long and of varying colors--some are white, others brown, and still others very dark. The young scale are called crawlers and only last for a few weeks after birth. The adult scale does most of the damage. They are found on the stems, foliage, and root parts of all woody and herbacious plants. Like the aphid, the scale attracts ants because of the honeydew. If you see ants, wasps, flies, or other insects around the plant, look close for scales. They do not move, or rather you cannot see them move. Some of the more harmful scale include: San Jose scale, oystershell scale, euonymus scale, rose scale, hemispherical scale, and tessellated scale.

The best method is to kill the scale with a dormant spray in the very late fall or early spring before the new growth begins. This is an oil compound that is found in all commercial plant stores or garden centers. Spraying with a dormant spray is a must for all commercial fruit tree growers, and should be part of your yearly control program. I spray in the spring and the fall for a double-barreled attack on a real pest.

SPIDER MITES

These are not really insects but relatives of the true spider. Let's just call them a pest. They are sucking insects and attack the tender parts of the plant such as new shoots, new buds, and new leaves. The damage on the plant appears in the form of stripped coloring on the leaves which will turn brown and lose normal color. The webbing is a way of telling if your plant has spider mite (red spider). If you look at spider mites under a lens you will see tiny spiders, green, red or yellowish.

Many species over-winter on the crowns and buds of perennials and weed plants. Some species over-winter in the egg stage on the bark of trees and begin their cycle in the spring. In closed conditions like greenhouses or indoor gardens, spider mites develop very fast and new generations may be seen In a very short time. The most common spider is the two-spotted spider mite. The cyclamen mite, broad mite, and spruce and bulb mite are also pests that do a lot of damage. Control of mites are all the same except the bulb mite, which I cover in the bulb section.

LEAF MINERS

These insects feed on the tissue between the upper and lower surfaces of the leaf. The result is a blotchlike or irregular serpentine mite. Typical miners include: boxwood miner, arborvitae leaf miner, and burdock leaf miner. These all produce a blotchlike appearance on the foliage.

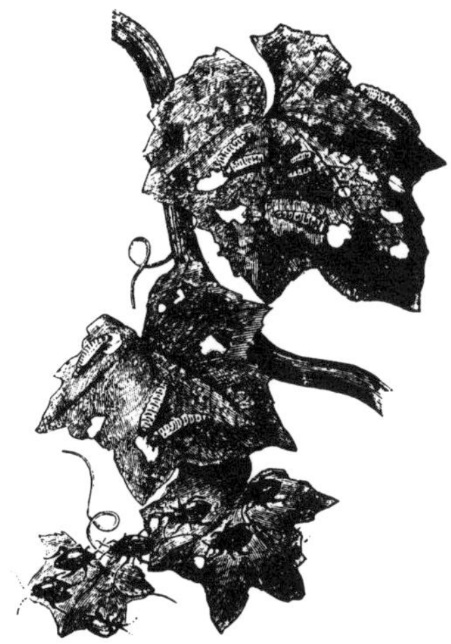

The columbine leaf miner and holly miners make serpentine mines on the foliage. Other forms of leaf miners (like the azalea leaf miner and the lilac leaf miner), after feeding on the interior of the leaf (mining the leaf), leave the mine and roll or tie the leaves together and use them as food. Others like the greenhouse leaf tier, sometimes called the celery leaf tier, have the habits as the azalea leaf miner but do not mine the interior of the new foliage.

The boxwood leaf miner is serious to many plants. They mine the interiors of leaves, leaving blotches or blisters on the surfaces. They are small yellow-orange maggots (larvae of the fly). When a heavy infestation appears on a plant the leaves will die and the plant will become disfigured. The greenhouse leaf tier is the common caterpillar type that rolls and folds the leaves. It may eat into the flowering parts of the buds as well as the interior of the leaf. The adult is a small tan-colored moth. It seems to be more of a greenhouse pest than an outdoor garden pest, but on occasion you will find it in the garden if the weather is humid and hot.

ALPINE BARRENWORT.

GALL-FORMING INSECTS

These insects sting the plant cells and stimulate plant growth, forming galls. These structures (see illustrations for various sizes and shapes) appear near the point of stimulation or attack. Among the most common are: spruce gall aphid, chrysanthemum gall midge, beaked willow gall midge, and dogwood club-gall midge.

Various Oak-galls.

[1] Solid galls on the cortex produced by *Aphilothrix Sieboldi.* [2] Bud-gall from a foliage-bud produced by *Cynips Hartigii.* [3] Solid galls on an Oak twig produced by *Cynips Kollari.* [4] One of these galls cut in half. [5] Bud-galls from foliage-buds produced by *Cynips lucida.* [6] One of these galls cut in half. [7] Leafy bud-galls produced by *Aphilothrix gemmæ.* [8] Bud-galls from foliage-buds produced by *Cynips polycera.* [9] Longitudinal section through one of these galls. [10] Gall on the pericarp of *Quercus pubescens* produced by *Cynips caput-medusæ.* [11–14] Spangle galls on a leaf of the Turkey Oak (*Quercus Cerris*); [11] produced by *Neuroterus lanuginosus*; [12] by *Neuroterus numismatis*; [13] by *Neuroterus fumipennis*; [14] by *Spathegaster tricolor.*

-Galls.

[1] Covering-galls on the petiole of the Black Poplar (*Populus nigra*) produced by *Pemphigus spirotheca*. [2] Scroll-galls on the leaves of an Alpine Rose (*Rhododendron ferrugineum*) produced by gall-mites. [3] Transverse section of one of these galls. [4] and [5] Bud-galls on the branchlets of the Wild Thyme (*Thymus Serpyllum*) produced by gall-mites. [6] Blister-like galls on the leaf of the Red Currant (*Ribes rubrum*) produced by *Myzus ribis*. [7] Part of the leaf seen from below. [8] Vertical section of a portion of this gall. [9] Solid gall on the leaf of the Gray Willow (*Salix incana*) produced by *Nematus pedunculi*. [10] The same gall cut open. [11] Part of the wall of this gall in vertical section. [1, 2, 4, 6,] and [9] natural size; [5] and [6] × 4; [3] and [7] × 8; [8] and [11] × 50.

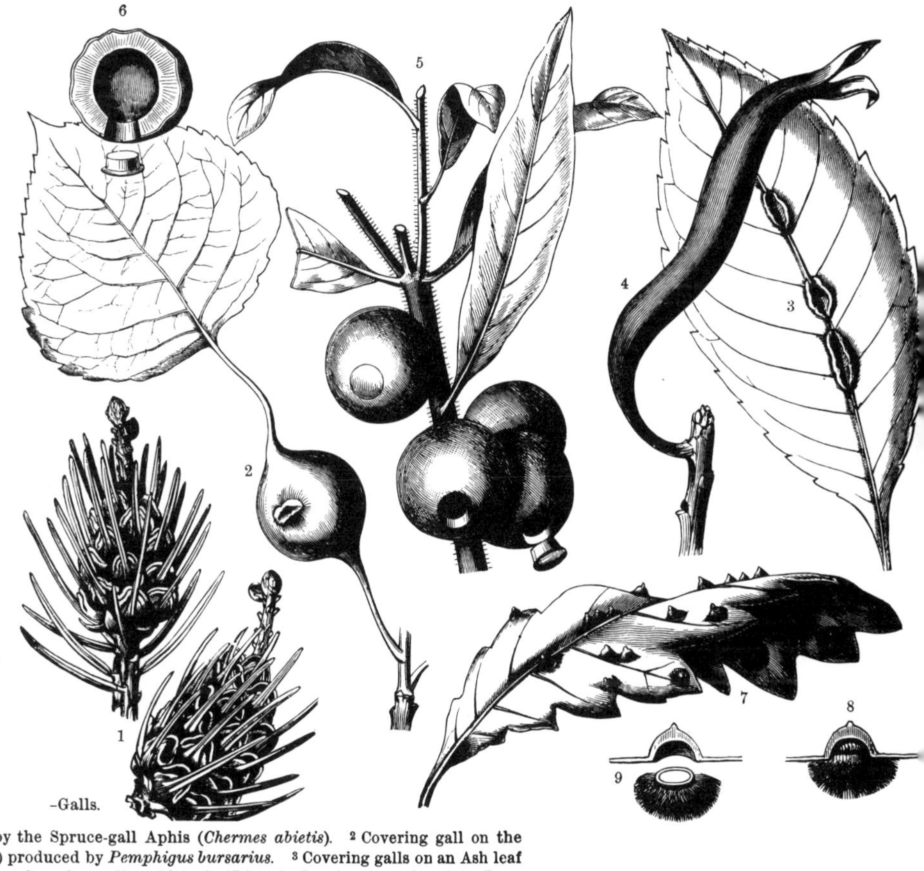

-Galls.

[1] Pine-apple gall on twigs of the Spruce Fir produced by the Spruce-gall Aphis (*Chermes abietis*). [2] Covering gall on the petiole of the pyramidal Poplar (*Populus pyramidalis*) produced by *Pemphigus bursarius*. [3] Covering galls on an Ash leaf (*Fraxinus excelsior*) produced by *Diplosis botularia*. [4] Covering gall on Pistacia (*Pistacia Lentiscus*) produced by *Pemphigus cornicularius*. [5] Solid galls on the cortex of *Duvaua longifolia* produced by *Cecidoses Eremite*. [6] Longitudinal section of one these galls. [7] Capsule galls on the leaf of the Turkey Oak (*Quercus Cerris*) produced by *Cecidomyia cerris*. [8] One of these galls cut through with the operculum still firmly attached, and [9] the same with the operculum falling away; × 3. The remaining figures natural size.

EASTERN SPRUCE GALL

The eastern spruce gall aphid causes cone-like growths on the new shoots of Norway spruce, white spruce, black spruce, and red spruce. The galls are about three-fourths of an inch long and look like small pineapples. The infested branches may die and the tree may become deformed and lose vigor.

This small blue-gray aphid, in the nymph stage, spends the long winter months on the twigs and branches of the spruce tree, usually at the base of the buds. In the spring they develop into adults, which are covered with a cottony secretion, and they lay eggs in groups. At the base of the new needles the young begin feeding, causing the growth of the galls. The insects are inside the galls. In the late summer the galls turn a brown color and open, and the adult aphid emerges. These aphids form wings and lay eggs on the needles which soon hatch. The young aphids over-winter on or near the buds.

COOLEY SPRUCE GALL

The Cooley spruce gall aphid causes galls on Engelmann spruce, blue spruce, and Sitka spruce on the East and West coasts. The only difference between the habits of this insect and the eastern spruce gall aphid is that the winged adults, upon leaving the galls in late summer, migrate to the Douglas fir where they lay their eggs and the young pass the winter. In the early spring the winged adults return to the spruce. Treatment is removing the galls from the plant.

CHRYSANTHEMUM GALL MIDGE

A common greenhouse gall infestation is the chrysanthemum gall midge. This is an orange-colored, two winged fly. The maggots are hatched on the tips of the new growth and are orange colored. At this point the maggots bore into the plant tissue. The galls are cone-shaped and develop at the point of attack. Both the larvae and the pupae complete their cycle in the gall. The complete cycle takes about a month and many generations appear in one season. If the infestation is heavy the flowers may not develop.

TIP AND STEM DAMAGING INSECTS

This group includes the larva and maggots of certain beetles, moths, and flies. These insects are very hard to detect until the damage is visible. They tunnel in the terminal shoots, the flower buds, and the main stem of the plant. Examples of this group of insects includes: pine tip and shoot moth, stalk borer, iris borer, lilac borer, dogwood twig borer, rhododendron borer, rose stem borer, and the flatheaded apple tree borer.

PINE TIP BORER

The pine tip borer moth, in the larva form, hollows out the tips, shoots, and the buds of pine trees. The presence of short, dead needles near the apex of the newly developed shoots, with partially devloped or hollowed out buds, is typical injury. If young trees are continually infested, year after year, they may become stunted and bushy. Tip moths are found in many locations in the United States and Canada. The Nantucket pine moth is common in the East and South, east of the Mississippi River. The larvae are yellow and about one-half an inch long. There is one generation in the colder regions and two in the semi-mild parts of the country and up to four in the humid Southern states. The insects over-winter in the pupae stage in the damaged tips of the pine. One form, found in the Northern Plains area over-winter in cocoons in the soil.

EUROPEAN PINE SHOOT BORER

Another borer that causes a lot of damage to the mugho pine and Scotch pine is the European pine shoot borer. Indications of infestations are the same as the pine tip borer moth.

STALK BORER

The stalk borer is the biggest offender of the flowering garden plants. They attack the stalks and fleshy parts of the aster, cosmos, dahlia, hollyhock, lily, peony, phlox, zinna and the goldenglow. This insect tunnels in the stems and is very hard to detect. You may not see the damage until the stem actually breaks or the plant wilts. A small hole in the stem indicates the point of entry of this pest and the castings are discharged here. The slender caterpillar is brown and a little over an inch long. If you split the stem you will find the pest.

The best treatment is the burning and removel of all old plants at the end of the season. This moth is grayish-brown and deposits its eggs in the late summer on weed plants like dockweed and ragweed for the next season's brood.

-Trailing Morning Glory—Convolvulus Mauritanicus.

IRIS BORER

The major pest of the irises are the iris borers. They attack both the Japanese and the Siberian types. The plant will turn black and decay. During the late summer the damage becomes more evident. This worm is pinkish with a brown head and is sometimes two inches long. The adults are brown moths that lay their eggs on the dead leaves at the base of the plant. The eggs hatch in the spring and the larvae begin the boring action. They work their way to the roots, which they will completely hollow out.

LILAC BORER

A major pest of woody shrubs like the lilac, privet, and many other flowering garden shrubs is the lilac borer. This pest tunnels under the back and into the wood. The stems may weaken and the foliage may wilt due to the girdling. This caterpillar is creamy white and about three-fourths of an inch long when full grown. It over-winters in the tunnels under the bark. The adult, a moth, emerges in the spring and lays its eggs on the rough bark or old wounds caused by the borer. You should cut off old branches that have indications of the infestation. When you see fine dust on the branches or bark, dig into the holes with a sharp knife. Commercial garden centers have a borer preparation that you squeeze into the holes before sealing the hole with a grafting compound or pruning paint.

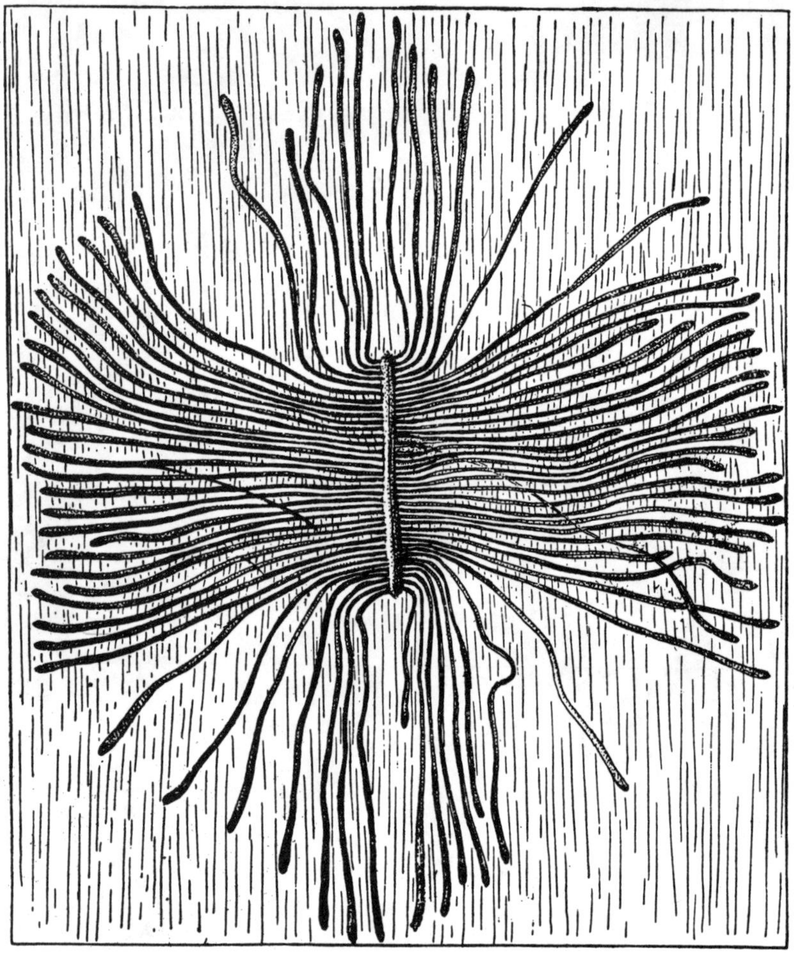

—Complete brood galleries of the hickory bark-beetle in surface of wood. (Original.)

—Exit holes in bark of hickory tree from which broods of the hickory bark-beetle have emerged. (Original.)

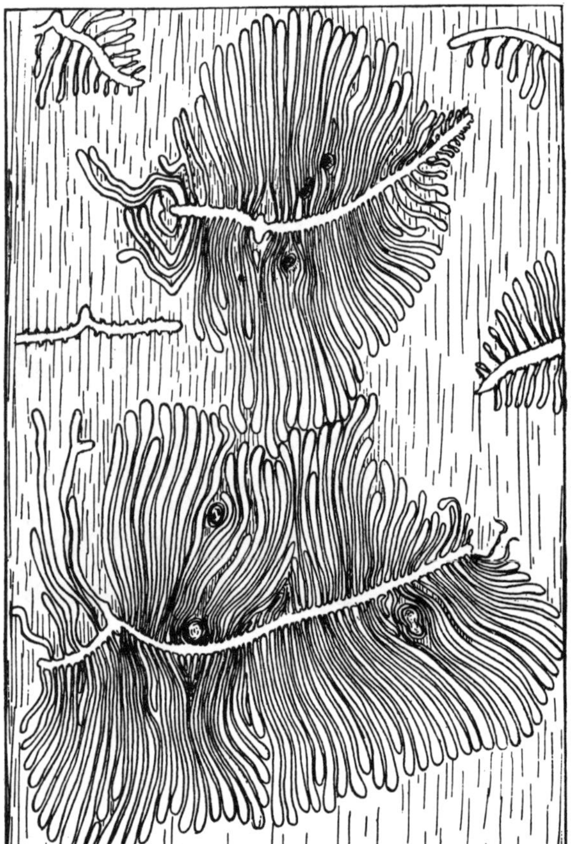

—Brood galleries of the cherry bark-beetle, showing character of primary galleries and larval mines. (Original.)

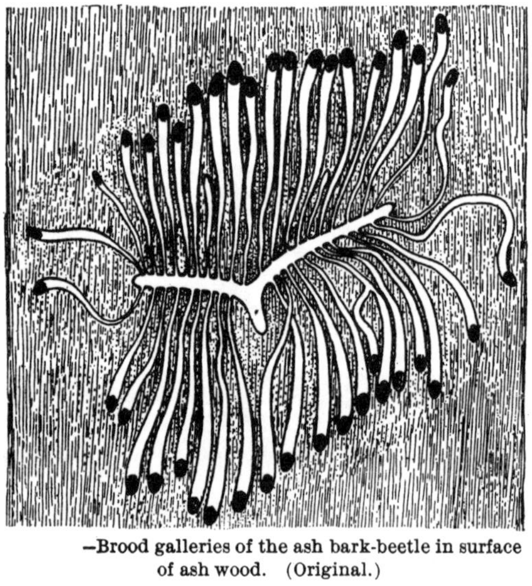

—Brood galleries of the ash bark-beetle in surface of ash wood. (Original.)

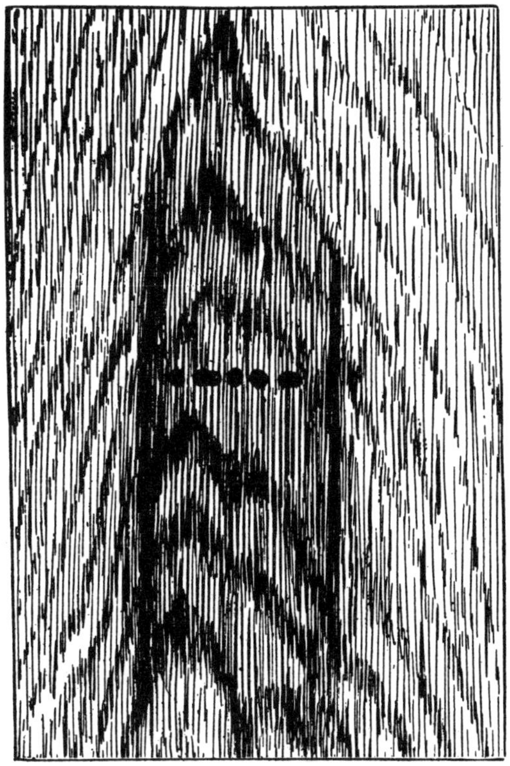

—Work of the Columbian timber-beetle: Black holes and "grease spots" in white oak. (Original.)

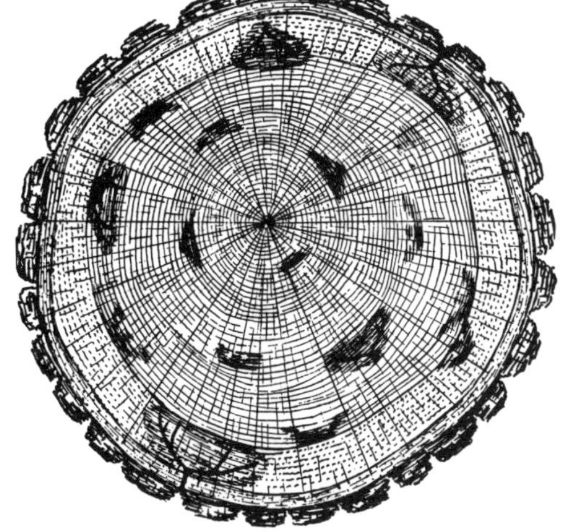

-Work of Columbian timber-beetle: End of log, showing evidence of serious damage, and how the brood galleries made in the sapwood are covered by subsequent growths ot wood.

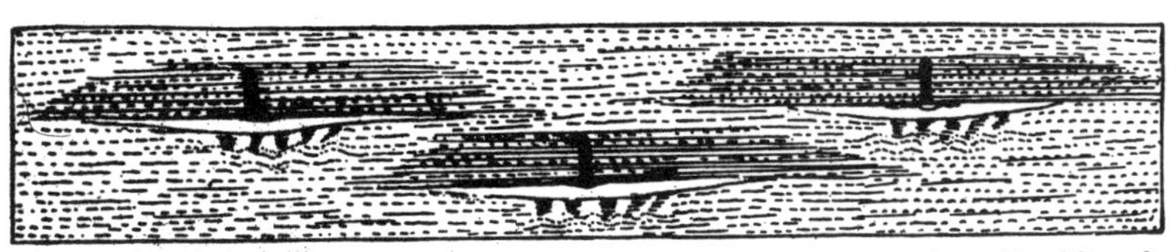

—Work of the Columbian timber-beetle: "Steamboats" in quartered or split white oak.

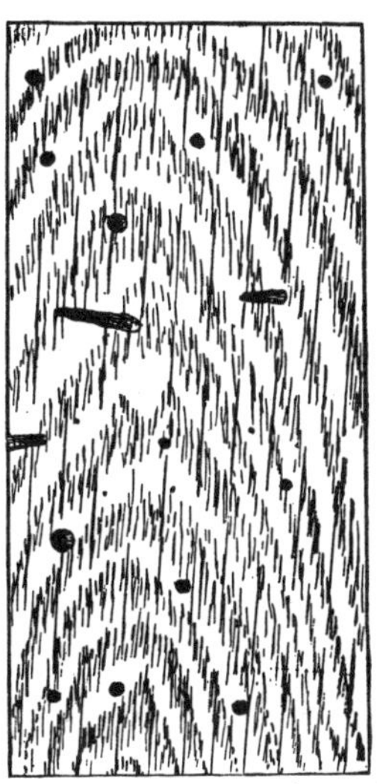

…—Pinholes in oak—work of the oak timber-worm.

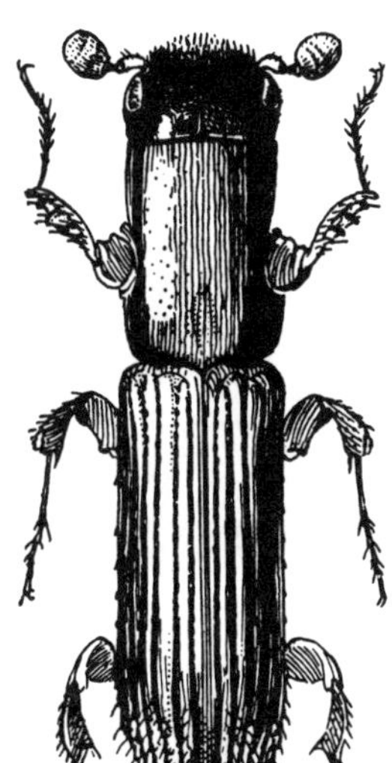

This cylindrical fungus-feeding beetle is admirably suited for living in round tunnels which it bores into forest trees for the propagation of its food.

-Wormholes in red oak—work of the oak carpenter-worm.

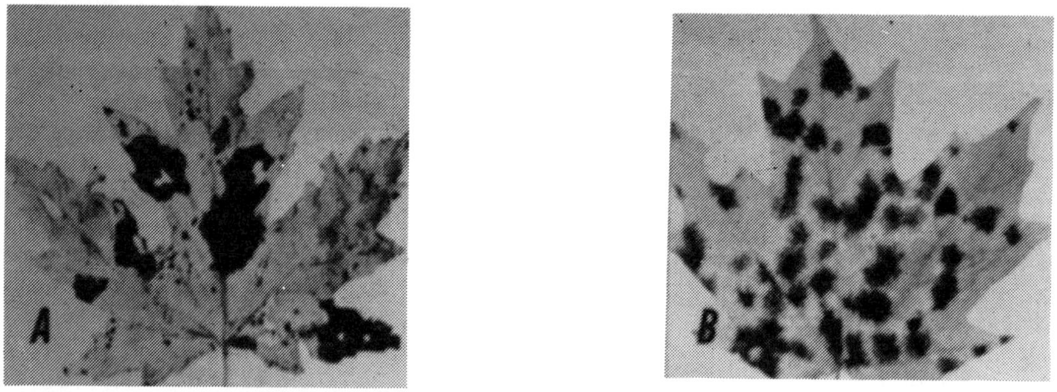

Maple leaves with tar spots caused by the fungi (A) Rhytisma acerinum and (B) R. punctatum

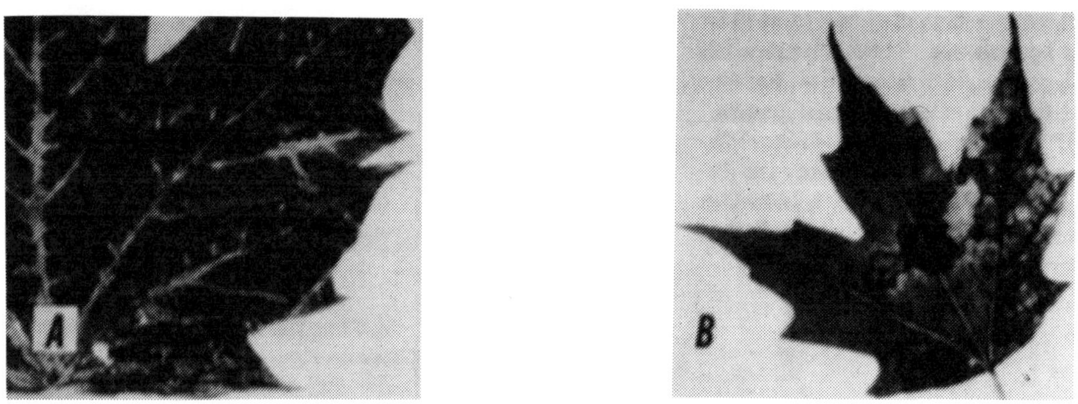

Leaves affected by anthracnose:
A. Norway maple
B. sugar maple

—Silver maple leaves eaten by larvæ of white-marked tussock moth in successive stages of growth from *a* (newly hatched larvæ) to *f* (full-grown larvæ)—reduced (original).

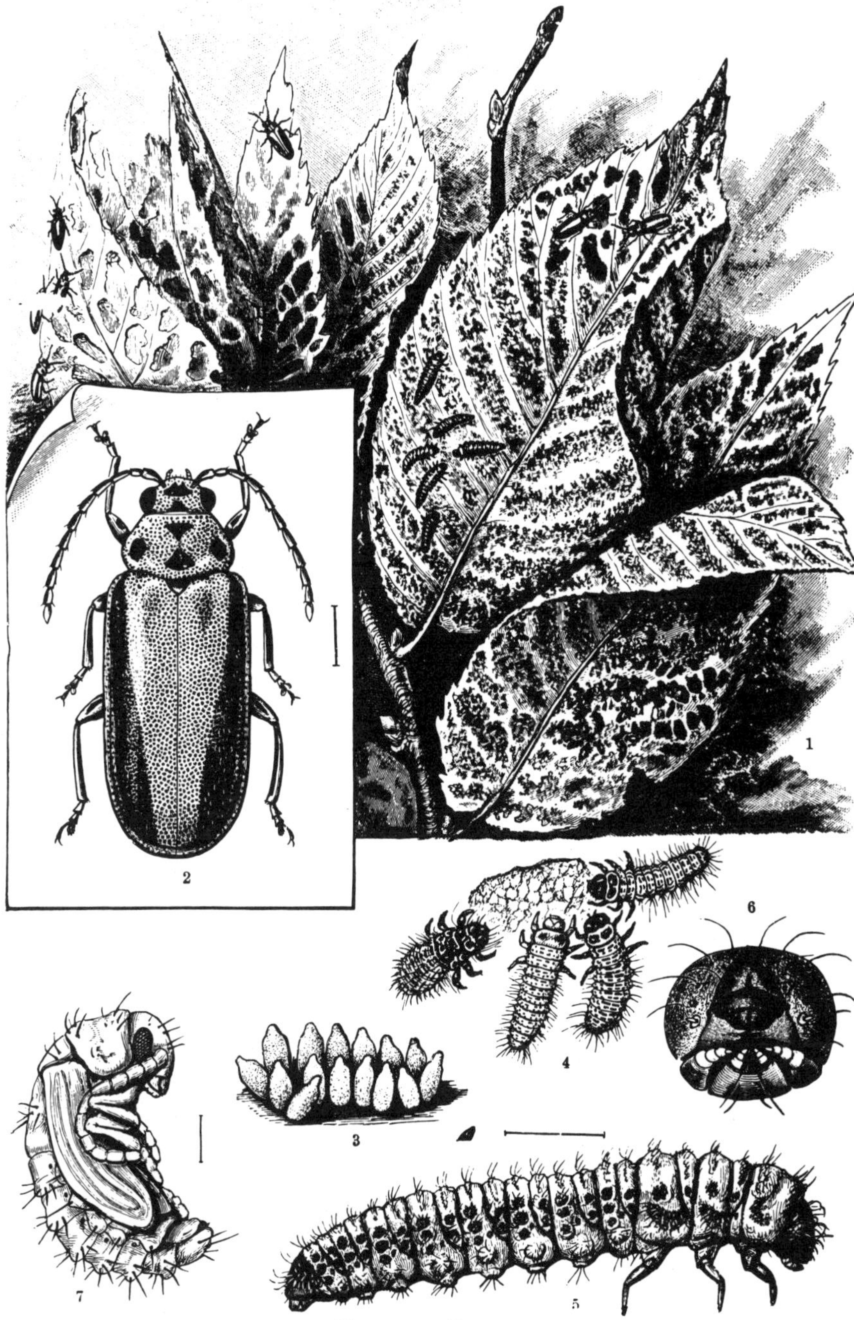

Elm leaf Beetle
(After Howard [Division Entomology], U. S. Department agriculture, Year book, 1895)

SOIL AND ROOT INSECTS

These crawling insects work in the soil and damage roots, tubers, bulbs, and corms of flowers. They feed on the surface and bore into the fleshy parts. Examples of the surface pests are slugs and snails, millipedes, and ants. Below the surface in the soil you will find the black vine weevil, strawberry root weevil, narcissus bulb fly, wireworms, and white grubs.

ANTS

Ants sometimes damage plants by tunneling around the base of the plant and root sections and sometimes will carry away seeds and seedlings. They also may be considered a pest because of colonizing and protecting aphids, scale, and mealybugs. They feed on the honey dew of these pests. They are also attracted by the sap flow of certain plants like peony and small ornamental flowering trees like the Japanese cherry.

TERMITES

Termites sometimes work in live plants by using the root systems as part of their tunneling home. If organic matter, or dead tissue of woody plants are not available, many times these subterranean termites will feed on living parts of the root systems.

MILLIPEDES, PILLBUGS, AND SNAILS

These are common garden pests and can be controlled by poison bait or just removing old dead plant matter that might attract them. They like damp places out of the light. Not a real problem to mature plants but will kill seedlings just emerging from the soil.

BLACK VINE WEEVIL

This is a snout beetle that attacks the stems and roots of many garden shrubs and herbacious plants. This pest attacks well over 75 kinds of plants found in the garden and greenhouse. The needles of the yew are bitten off at the top along one side or are eaten completely by the adult. The larva are white and feed on the roots and will strip and bark from the larger roots. They attack maiden hair fern, gloxinia, arborvitae, hemlock, rhododendro andromedia, primrose, tuberous rooted begonias, and wisteria.

Outdoors the insect breeds on strawberry, yew and rhododendron. The adult is black with patches of yellow hair and is about two-fifths of an inch long. They over-winter in the larva or pupae stage and the wingless adult emerges in the late spring or early summer.

PEONY BLIGHT

This seems to be the biggest problem for peony growers. The "blight" is a fungus called botrytis and is common in all parts of the country. The symptoms of the fungus are a crown rot and, on new plants, the shoots become black. This is the first phase of the disease. In the second phase (caused by another species of the same fungus that affects not only peonies but many other varieties of garden flowers), it turns the flower buds black and the flower pedals and leaves become spotted. Control programs are very successful and include:

1. Removing all old stock in the fall.

2. Hand picking wilted new shoots in the early part of the growing season.

3. Picking off wilted flowers.

4. If you use chemicals in your garden, you can spray the young shoots in the early spring and spray again before the buds develop. The final part of the spraying program is in midsummer to control the leaf blight part of the cycle. The type of chemical you use depends on your preference, but most of the commerical fungicides will do the job.

FIREBLIGHT OF ORNAMENTALS

YELLOW JAPANESE KERRIA.
(KERRIA JAPONICA.)

This disease attacks many trees and shrubs and does a lot of damage. Many times this disease is associated with one or two kinds of trees or shrubs and is labeled such, but it is a commom ornamental pest and here are the symptoms:

1. The foliage near the terminal tips turn brown and black and wilt.

2. The foliage looks as if it had been scorched by fire.

3. On large trees, large cankers can develop on the trunk and the branches.

There is little that can be done to prevent this disease, but you may help by cutting out the diseased parts of the plant. It is a disease of the interior of the plant and a fungicide will not help.

PESTS OF ROSES

MRS. R. G. SHERMAN CRAWFORD.

ROSES

PESTS OF ROSES

Insects which most commonly affect roses are of two general types: those which eat the foliage, as rose slugs and rose chafer, and those which suck the sap, as aphids, scale, and thrips.

The presence of leaf-eating insects is usually first detected through the discovery of partially eaten leaves or of skeletonized leaves, that is, leaves from which a portion of the lower or upper surface has been eaten, leaving the other surface as a transparent membrane, or with the fleshy part of the leaf completely eaten and only the veins and the stem left. The only protection is handpicking the insect or spraying with a chemical or organic insecticide.

Sucking insects obtain their food by sucking the juice out of the stems, buds, and the leaves of the rose. Aphids are usually on the youngest growth at the tips of the branches and on the underside of the leaves. Thrips injure the flowers, while the scale insects usually inhabit the woody parts of the plant.

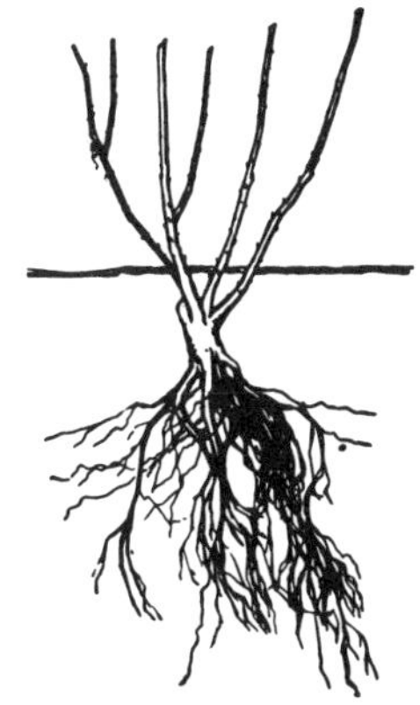

Proper Depth to Plant

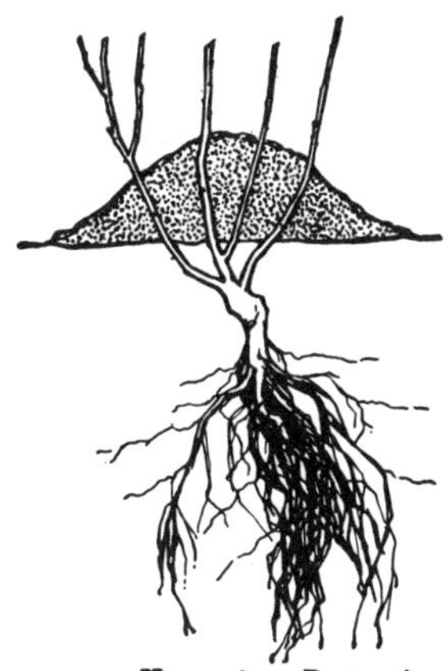

November Protection

FUNGUS PARASITES

Roses are very rarely killed outright by fungus attacks, but by their presence the vigor of the plants is greatly reduced, the foliage may be rendered unsightly, branches become more or less distorted or disfigured, and the quality of the blossoms lowered. In considering the fungus diseases of these plants, it must be recognized that (in common with similar diseases of other plants) treatments are preventative rather than curative. Thorough and repeated sprayings with a suitable fungicide will in most cases so check the development of the parasite that the disease will be in great measure controlled for the growing season. However, the full benefit of fungicidal applications can only be secured by their early use in the spring, guarding in this way against the recurrence

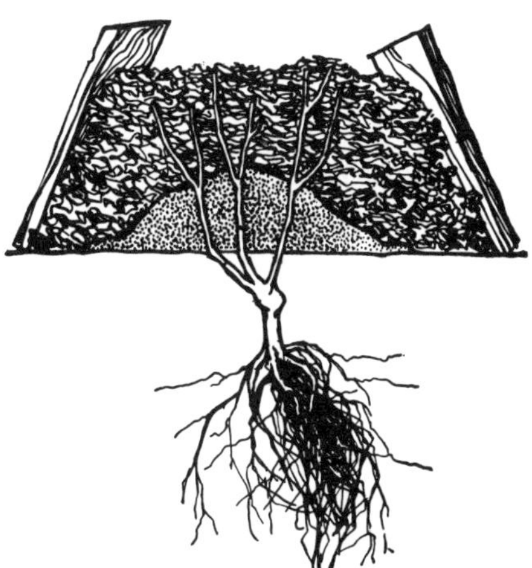

December Protection

of a disease of the previous season. Fungicides to be recommended for rose diseases are Bordeaux mixtures.

POWDERY MILDEW

Powdery mildew is extremely common, few gardens being entirely free from this most destructive of all rose diseases. Wild as well as cultivated roses suffer from it, crimson ramblers being especially susceptible. There is, however, a wide difference in the susceptibility of different varieties, as well as of the same variety in different localities. Each rose grower should give special attention to the degrees of resistance shown by various sorts in his collection, keeping only those which show the greatest resistance to the disease.

The fungus frequently appears very early in the season on young buds, leaves, and young shoots. It causes the appearance of a delicate white mildew that becomes mealy after the development of the conidia, or summer spores (microscopic bodies which serve to reproduce the fungus). If the attack is severe, the normal development of the succulent young shoots is arrested, the leaves become curled and deformed, falling prematurely, and complete defoliation sometimes results. The fungus mycelium (microscopic thread-like growth of the fungus) forms, later in the season, feltlike covering on the twigs, calyx, fruit, and even on canes. Minute spherical black bodies in which the winter spores develop can be seen on the felt. These spores are capable of persisting through the winter and setting up a new infestation in the spring.

The disease may be held under control during the stage in which the summer spores are produced in immense numbers. Several treatments with a fungicide, applied very thoroughly at intervals of 10 days to two weeks should do it. If the development of the parasite has not been completely arrested in its summer stage and if the winter spores

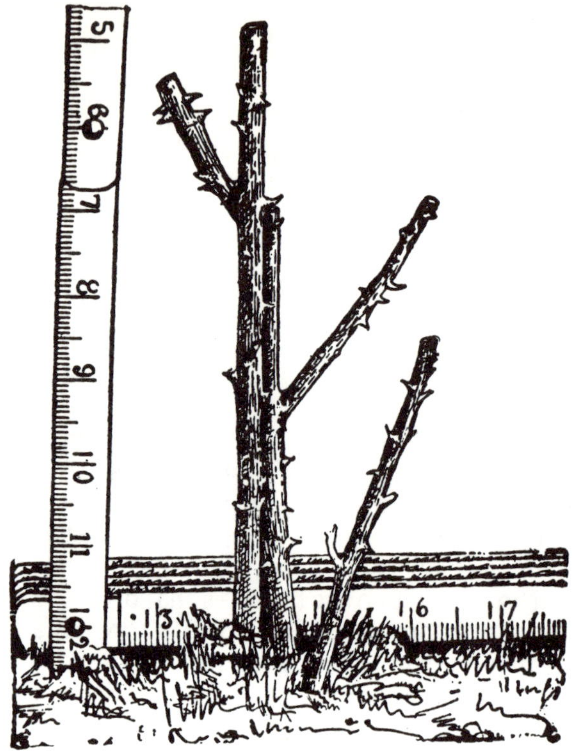

.—A hybrid tea rose (Radiance) after pruning for individual blooms.

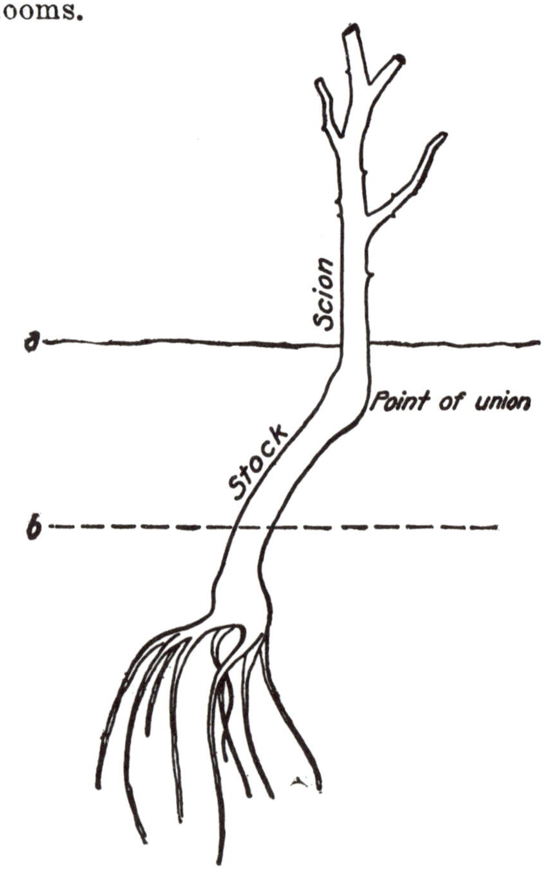

–Proper depths for planting roses: *a*, Proper soil level for a grafted or budded rose; *b*, proper soil level if the plant were on its own roots.

have been permitted to develop, more energetic measures must be taken. Late in the fall diseased portions of the bushes should be trimmed off and removed from the garden site, together with all fallen leaves and debris that have collected under the shrubs. The plants should then be thoroughly sprayed with Bordeaux mixture, as recommended for dormant plants.

ROSE RUSTS

The disease known as rusts are caused by any one of several parasitic fungi, and they occur on both wild and cultivated bushes. Any part of the plant may suffer from the presence of the fungus, but the effect varies with the time of the year and the part attacked. On the green parts of the bush in the spring or early summer, conspicuous orange-colored swellings appear that later in the season take on a deeper shade and become powdery. On the leaves the effect may be shown by small circular spots, but when the swellings occur on the young green stems and petioles, those parts become bent and greatly distorted. During these stages of growth the parasite has borne two kinds of spores, but later a third kind, the winter spores, develop either in small black groups on the underside of the leaf or in more-or-less elongated black powdery patches on the woody stems.

The spread of the disease in its early stages can be checked by using a commercially prepared treatment, but it is advisable to cut off and destroy the diseased parts of the plant. Late in the season, after the flowers all have been cut, I would completely cut the plant back for winter mounding and remove all the plant parts. At this time it is a good idea to give the plant a good spraying with Bordeaux mix. You can buy this in most complete plant stores. If after a few seasons you cannot get rid of the rust, for rust can over-winter on healthy plants, I would replace the infested plant. Better to destroy one plant than have a rose garden of rust.

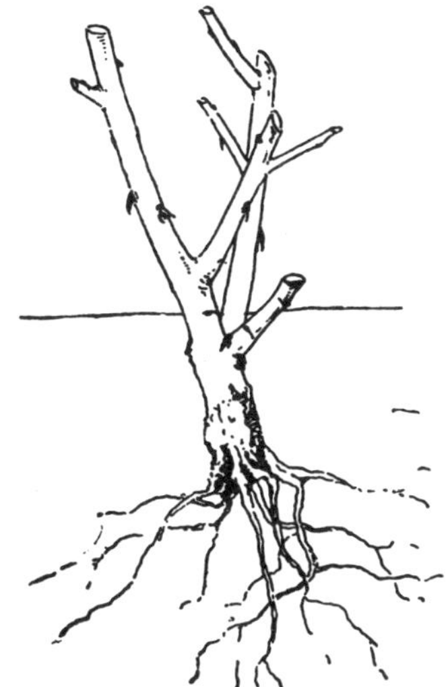

TOO DEEP

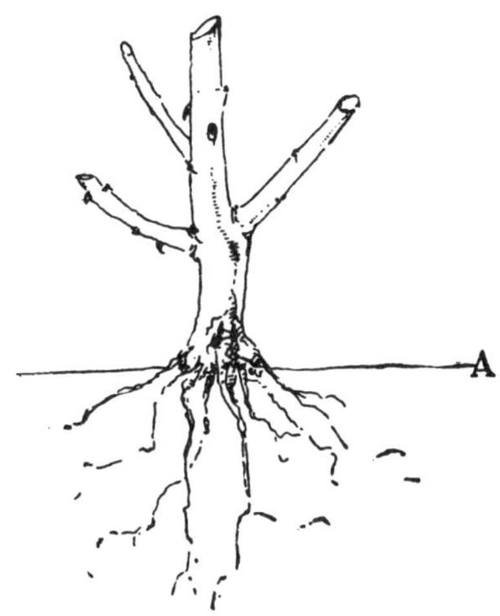

TOO SHALLOW

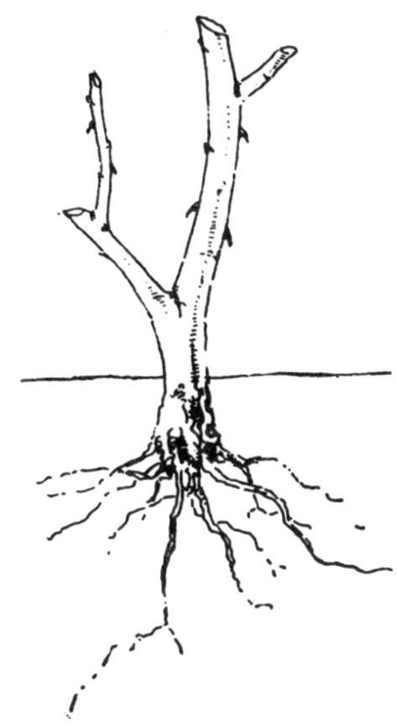

PROPER DEPTH

ANTHRACNOSE

The new growth is the most susceptible to this fungus. On the leaves blotches appear that contan the fruiting bodies. These later become pinkish from the escaping spores. Upon the surface of the branches similar fruiting bodies may entirely surround the stem and spread to the tip of the plant. The foliage on such plants become pale and dwarfed and will soon fall. If the attack is severe, the plant may completely defoliate. As in rust, the best cure is prevention and removel of all diseased parts.

Rose root gall

LEAF SPOT

The leaves develop a brown center with a purple border. There are several fungus that cause spotting of roses but all have the same symptoms. Pick off leaves that are spotted before the fungus can spread.

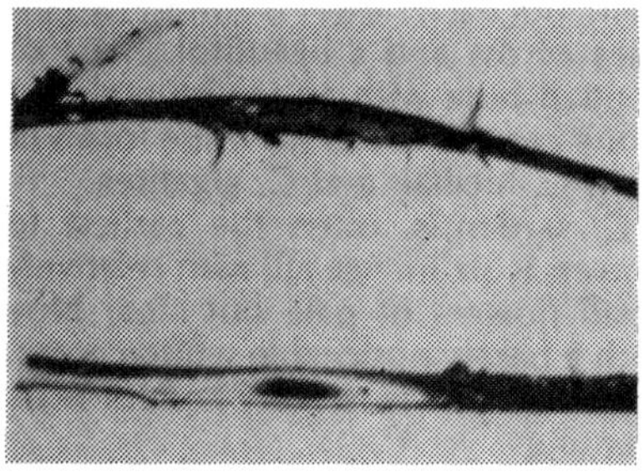

Rose interior gall

LEAF BLOTCH

This is the most serious of all the rose problems. It is commonly called "black spot" and probably is found in every rose garden in the world. Today they have developed disease-resistant varieties and it would be to your benefit to grow them. I think there must be hundreds of disease resistant roses of different colors and different shapes, so you can pick one to fit your personality or color to match your house.

The first symptoms are irregular black spots on the foliage. The spots soon spread to the tender new stems as well as other leaves. They increase in size and may spread to the whole plant.

The second stage of this disease matures upon the fallen foliage, and the spores living over the winter are able to set up a new infestation. This shows the real neccessity of cleaning up the rose garden after the growing season. If the old leaves and tops of the plants are cleared out, the spores have no place to over-winter and you will have succeeded in eliminating most of the trouble. In the spring, spray the plants with a fungicide before the fruiting bodies appear on the plant. A little prevention never hurt the garden.

Rose globular gall

COMMON CANKER

Outdoor roses in many locations suffer severely from two forms of canker not always readily distinguished by the ordinary home gardener. The early symptoms of a disease long recognized as a canker is the appearance of reddish patches on the green parts of the plant, generally on the first year's growth. The disease is caused by a fungus classified as a wound parasite. That is, the spores gain entrance to the shrub through certain mechanical difficulties such as broken stems or bruised branches. The break can be as small as a thorn rubbing against another shrub, or by insect punctures. The only way to get rid of cankers is to cut the canker off the plant. You can spray with a fungicide but cutting off the canker is still the best method of cleaning up the plant. There are roses that are not susceptible to canker and I suggest you try them. Your rose catalogue usually tells you the best region and condition to grow certain roses in, and the varieties that are resistant to diseases.

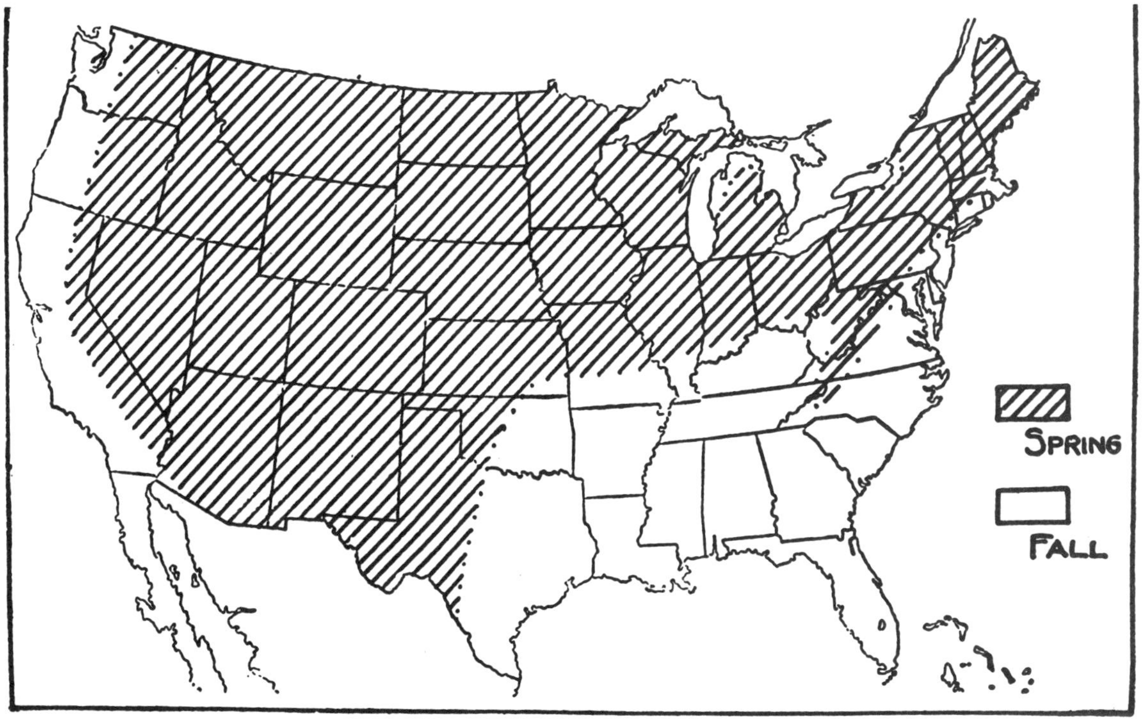

—Outline map of the United States, showing the areas in which fall transplantings and spring transplantings of roses are usually most satisfactory.

PESTS OF SUMMER AND FALL BULBS

FLOWERING BULBS

To properly identify and control diseases of bulbs I have divided them into three categories:

. Hyacinths and Tulips

. Daffodils and Narcissus

. Gladiolus, Iris, and Lilies

RULES FOR HEALTHY BULB DEVELOPMENT

1. Buy from a reliable nursery or garden center and always select number one grades. Make sure the bulbs are hard. The bigger the better.

2. Plant in a well-drained soil and fertilize with bonemeal.

3. Never plant indoor bulbs in garden soil. Clean out the pots in hot water and use bagged soil with gravel for drainage (for forcing bulbs). If you bring in soil and stones from the garden you may bring in diseases and insects that live in the soil.

4. Plant the bulbs at the proper time and proper depth. The bulb chart will clearly outline the various depths of different bulbs.

GROUP OF NARCISSUS.

Hyacinth Bulb with Offsets

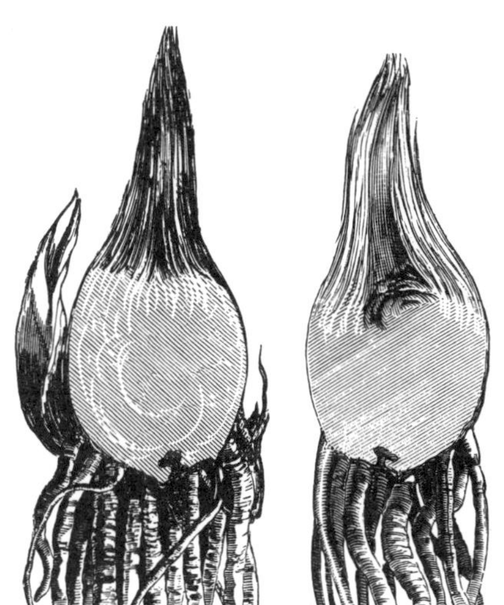

—SOUND BULB. —BULB DECAYED AT CENTER.

PESTS OF HYACINTHS AND TULIPS

FIRE (BOTRYTIS)

Bulbs: Fire forms a black fungus on the husk of the bulb.
Foliage and tender shoots: Foliage and stems are twisted and rotten. Leaves have yellow-brown spots and usually are covered with a mold.
Flowers: Often the flowers do not open or have spots like the leaves.
Control: If there is any evidence of black sclertia at planting time in the fall, remove the husks, spray, and continue to spray after plants begin to show through the soil. Always destroy diseased parts of the plant after flowering.

BULB CROWN AND ROOT ROT (FUNGI AND BACTERIA)

Bulbs: Look over the bulbs at planting time and throw away all bulbs with a musty odor and a dry chalky appearance. The bacterial rot is soft, musty, and bad smelling.
Control: Eliminate all bulbs with above appearance. Never plant a soft bulb. Dust the bulbs at planting with a fungicide. If a bulb fails to come up in the spring, remove it from the soil and take some of the soil from around it. Avoid heavy nitrogen fertilizers in bulb beds. Do not over-water.

FLOWERING BREAKING AND MOSIAC

Bulb: The bulb is usually smaller than normal but no outward symtoms.
Foliage: Dark and light green mosiac.
Flowers: Streaks of white or dark color depending on the strain of virus.
Control: Some varieties of tulips are naturally infected with virus. The sellers of bulbs, both wholesale and retail, usually have lots of information on the best variety for your soil condition and location. If you write to the Dutch Bulb Growers (you can get their address on any bulb package in your garden center), they will supply you with a list of all bulbs with natural viruses.

I would not plant any solid colored bulbs next to any strain that naturally has virus. Many viruses are spread by aphids, so keep the insect population under control. There is no control for this virus except to rid the garden of all plants that have the infection.

SUNSCALD OF TULIP

Bulbs: They rot.
Foliage: Twisted and abnormal growth.
Flowers: Petals dry out at upper margin.
Control: Sunscald is caused by tulips being planted in a very southern exposure. When the snow melts in the spring, and the soil gets dry too early, the plants lose a lot of moisture. Add peat-moss to the existing soil to retain moisture if you see the water is draining off too fast in the beds. Keep the bulbs well watered in the fall to replace moisture lost by the foliage.

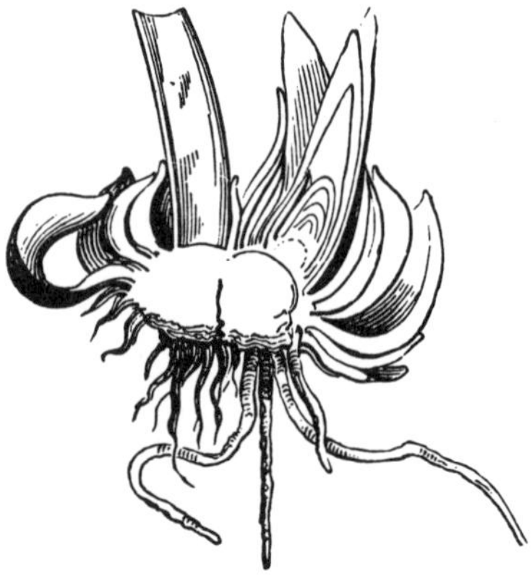

Lily—Scaly Bulb

WINTER INJURY

If the young shoots come up in the spring twisted or abnormal, and no flower appears, the bulb has been damaged by winter injury. The bulb probably has been rotted by poor drainage in the bulb bed or heavy soil. Try to plant bulbs in a loose-mix soil, early enough in the fall to allow good root systems to develop. The bulb chart will give you proper depth for each type of bulb. To improve water drainage, punch holes in the beds with a pipe or sharp instrument. If the soil is heavy, add peatmoss and sand, soil conditioners like vermiculite or perilite (both are inert materials to lighten the soil), and cow manure and bonemeal. Bonemeal builds roots and should be included in the holes in the fall when you plant the bulbs. It may help to cover the beds in the fall with peatmoss or salt hay. This will keep the soil temperature even during the temperature changes in the colder months--it sort of keeps them in deep freeze.

Every few years you should dig up the beds and separate the old bulbs from the new bulblets. Throw out the old bulbs and dust the new bulbs with a fungicide. If you keep the beds clear of weeds during the brief period the bulbs are in bloom, you will help keep the insect population in check. If you can remember where you planted every bulb, you can cultivate early in the spring, as soon as the soil is workable. (My memory at this stage in my life is nowhere near that ability. I am lucky to remember to plant bulbs in the fall.)

FROST INJURY

The bulbs are not affected by frost, but the new growth, just breaking through the soil can be severely damaged by a spring frost. Tulips are not as susceptible to frost-kill as hyacinths, but should be covered with a plastic cloth or a wooden box if frost is forecast. Anything that does not touch the plant and keeps the frost off foliage will work.

PESTS OF DAFFODILS AND NARCISSUS

BUD BLAST

Bulbs: No external symptoms.
Flowers: Buds turn dark and brown and dry up.
Foliage: No external symptoms
Control: Narcissus need good root development.

BULB ROT (FORCED BULBS INDOORS)

Bulb: Bulbs are soft and rotten.
Foliage: Leaves become twisted and abnormal.
Flowers: None.
Cause and Control: This condition is caused by a variety of gardening practices.

1. Over watering the potted plants.
2. Poor or no drainage in the pots.
3. Not enough time in the potting shed or cool room to develop roots.
4. Drastic temperature changes from the cool room (where the roots develop) to the low humidity and high temperature of the home. When you force bulbs indoors you must try to simulate the exact conditions and general time it takes the plant to set roots and grow in the normal outdoor environment.

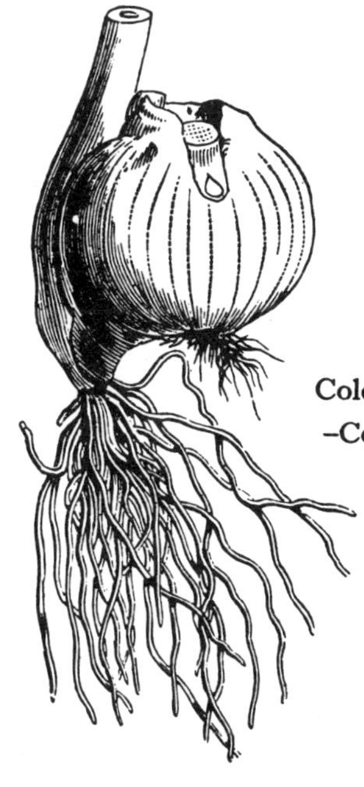

Colchicum –Corm

CROWN AND ROOT ROT (SEE TULIP)

FIRE

Bulbs: A black fungus forms on the husk of the bulb.
Foliage: A yellowish-brown spot appears on the leaves. The new growth comes up in the spring twisted and deformed. The spots are usually surrounded by a circle of water.
Control: Low humidity is a natural control. This disease usually occurs in the greenhouse or forcing frame. Open the vents and let in some air if the humidity gets too high.

FLOWER BREAK (VIRUS)

Bulb: No external signs or symptoms.
Foliage: No outward symptoms.
Flowers: White streaks on petals.
Control: None (you have to remove and destroy all the diseased parts).

FLOWER STALK COLLAPSE

Bulbs: No external symptoms.
Foliage: The stalks develop water soaked regions and collapse.
Flowers: None.
Control: Too fast a temperature change. Gradually change the temperature when moving potted bulbs indoors from the cool room. First set the bulbs in a room that is forty to fifty degrees, then move them into a slightly heated room, and finally into the house.

MOSAIC (VIRUS)

Bulb: No external symptoms.
Foliage: Mottled with light and dark green spots.
Flowers: Sometimes a dark streak but other times no external symptoms.
Control: None (you have to remove the diseased plant parts).

PAPER TIP AND WHITE STREAK

Bulb: No external symptoms except at times the affected bulbs are smaller than grade "A" bulbs.
Foliage: The leaves usually wilt and die.
Flowers: The flowering tips dry out and become brittle.
Control: Control the insect carriers in the garden and you will help control this disease. There is no known cure for paper tip and white streak.

TYPES OF GLADIOLUS.

PESTS OF GLADIOLUS, IRIS, AND LILIES

GLADIOLUS THRIPS

The gladioli, iris and lilies are affected by this insect. The full page illustration clearly shows the life cycle and damage done to the flower and green parts of the plant. The life cycle of the gladioli thrip begins when the adult lays the eggs in the tissue of the growing plant. The eggs hatch into small yellow nymphs. The total life cycle, under ideal conditions, takes only a few weeks.
Damage: The leaves take on a silvery color, then turn brown. The flowers of the gladioli become deformed and spots appear. When the infestation is severe the flowers do not develop at all. The adult thrip over-winters on the corm of the gladioli, so precautions should be made to provide a clean storage area with proper temperature and humidity controls. Dust the corms with an insecticide before storing. Buy corms only from known nurseries or bulb sellers. If you buy in the spring, dust the corms before you plant them in the late spring or early summer. If you are interested in growing glads for show or just want more information about this very popular flower, I suggest you contact the Gladioli Society. It publishes a yearbook with all

GLADIOLUS CORM WITH FULL GROWN OFFSETS.

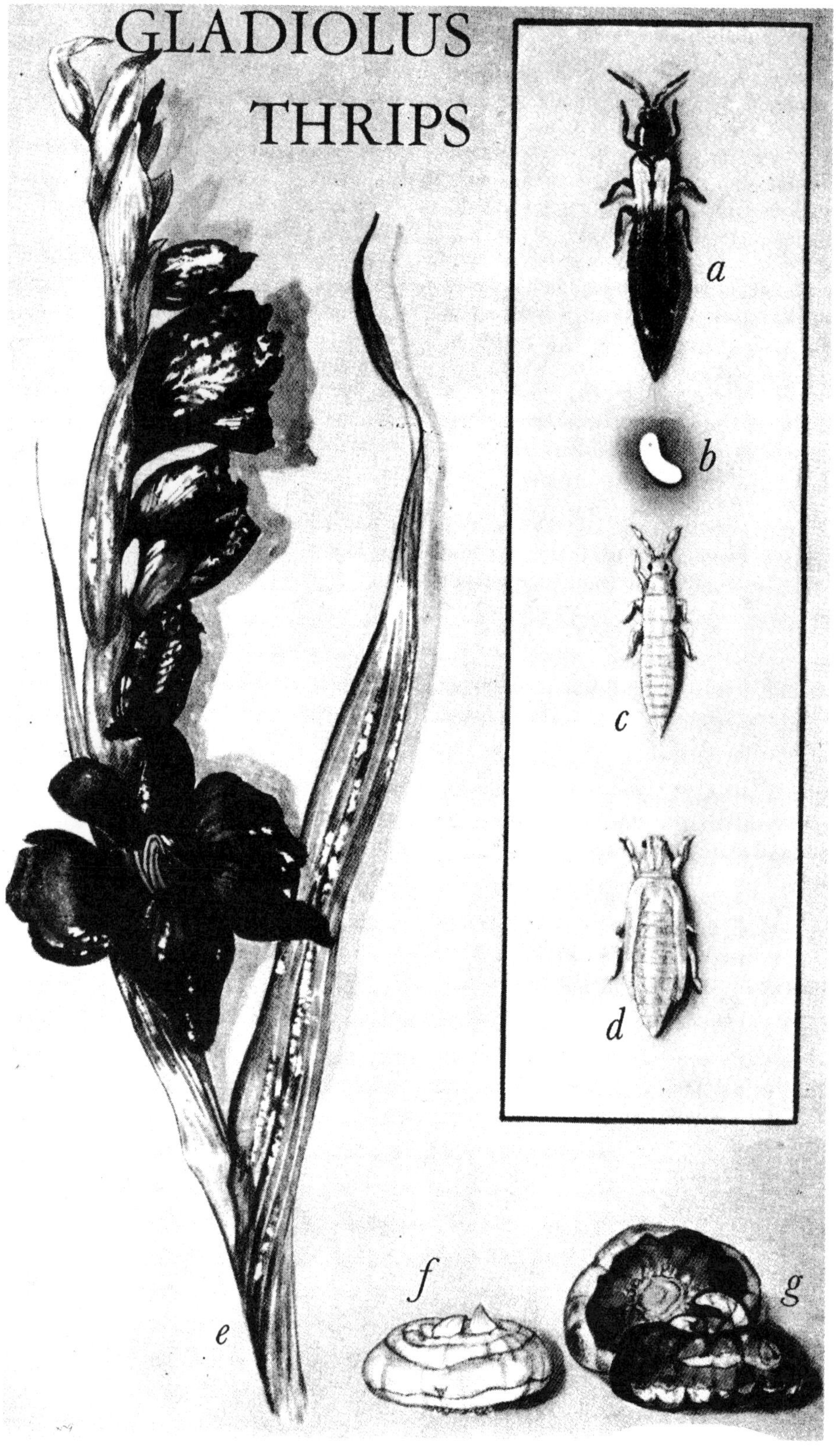

a. adult thrips; b. egg; c. larva; d. pupa; e. foliage and flower spike showing typical feeding injury; f. uninjured gladiolus corm; g. corms injured by feeding of thrips, showing characteristic russeted appearance; (all enlarged greatly).

the new information about controls and new disease resistant varieties.

GLADIOLI DISEASES

There seems to be only three main diseases affecting glads. They are corm rot, viruses, and blight. (Sometimes gardeners mistake chlorosis (lack of available iron in the soil) as a disease, but this can be overcome with proper fertilizing.)

BLIGHT

Blight develops during wet weather. It affects both the foliage and the flowers. The symptoms are yellowing of all plant parts, wilting, complete kill back, and the foliage will have spots. To control this disease it is advisable to spray with a fungicide once or twice a week during the moist spring season. Do not spray until the plants have reached a height of ten inches. After you cut the flowers, remove all parts and destroy.

CHLOROSIS

This is not a disease but does affect gladilus and is very common. Many times this is caused by too alkaline a soil. The Ph of the soil should be about 6.0 and slightly acid, to control any yellowing that is related to chlorosis.

CORM ROT

Of all the diseases affecting the gladioli, corm rot is by far the most serious. Its symptoms are sunken, corky lesions on the corm, usually caused by rough handling during storage. The storage and care is the key to successful gladiolus. The temperature at the time of harvest should be about seventy-five degrees for about fifteen days. This will eliminate all the cuts on the corm. The tempera-

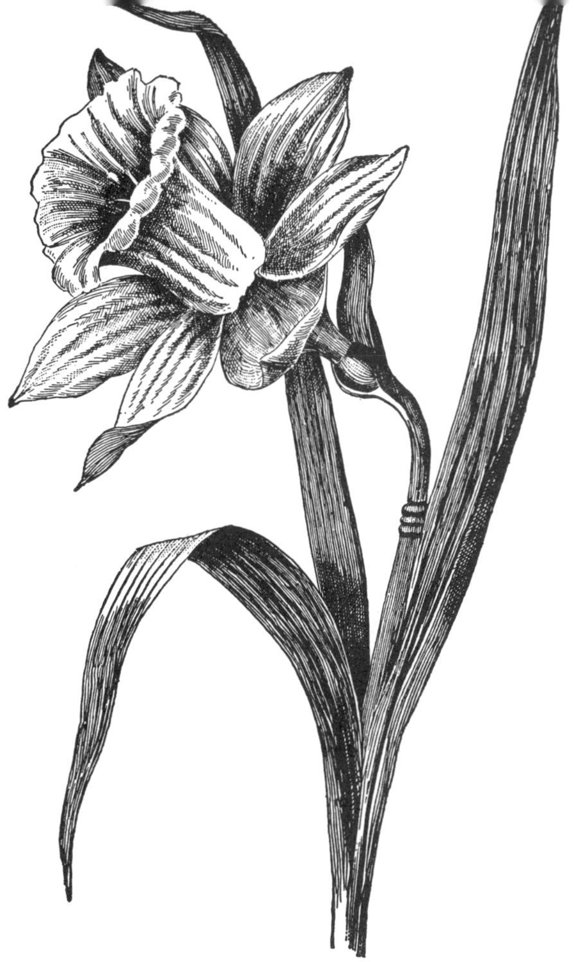

TYPE OF HOOP-PETTICOAT NARCISSUS.

LILY OF THE VALLEY.

ture should be between thirty-five and forty degrees in storage. If you pre-dip the corms in a fungicide at the rate of two teaspoons per gal. of water, this will help reduce the disease.

VIRUS

The most common gladioli virus is called white break. The symptoms are white streaks on both the foliage and the flowers. If the plant is infected before the blooming period begins it may fail to bear flowers. To control viruses you must remove the diseased parts of the plant. Before you cut flowers or just prune, make sure the shears have been steamed or well cleaned. Keep the beds free of weed plants for insects carry diseases and are attracted to weedy gardens. Good cultivation and careful practices are the best and sometimes only control of virus diseases.

POTTS' MONTBRETIA OR TRITONIA.

YEARLY CONTROL PROGRAM FOR GLADIOLUS

1. Do not bruise the corms when storing. Follow temperature levels that were mentioned before.

2. Only plant healthy, large corms.

3. Pre-dip corms in a fungicide.

4. Set up a spraying program that includes both fungicides and insecticides (if you use chemicals). You can control a lot of the problems by following common, natural practices, like companion planting to keep down the aphids that carry diseases.

5. Keep the weeds down.

6. Cut off tops of plants after flowering time and remove from garden site.

7. Try not to use the same bed for gladiolus more than two years in a row. This will prevent build-up of diseases and other gladiolus pests.

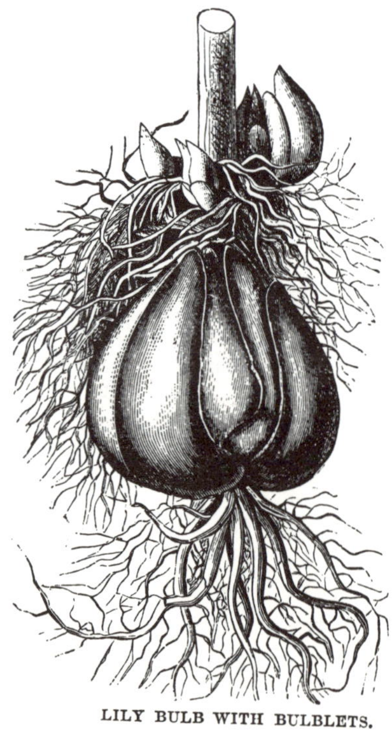

LILY BULB WITH BULBLETS.

MINOR BULB PLANTING GUIDE

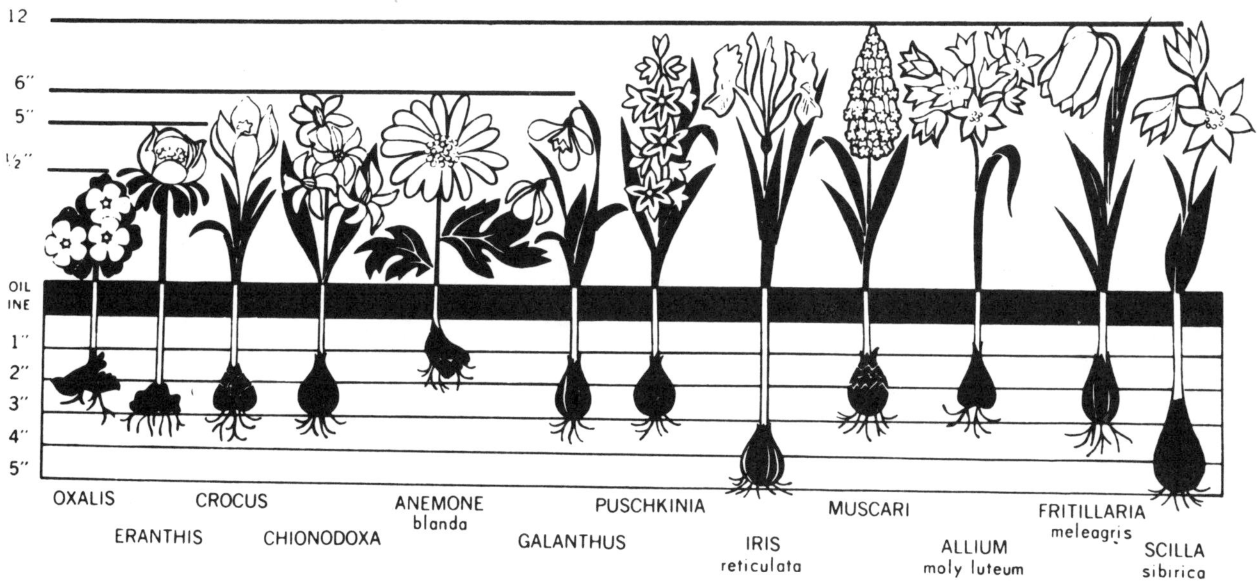

TULIP PLANTING GUIDE

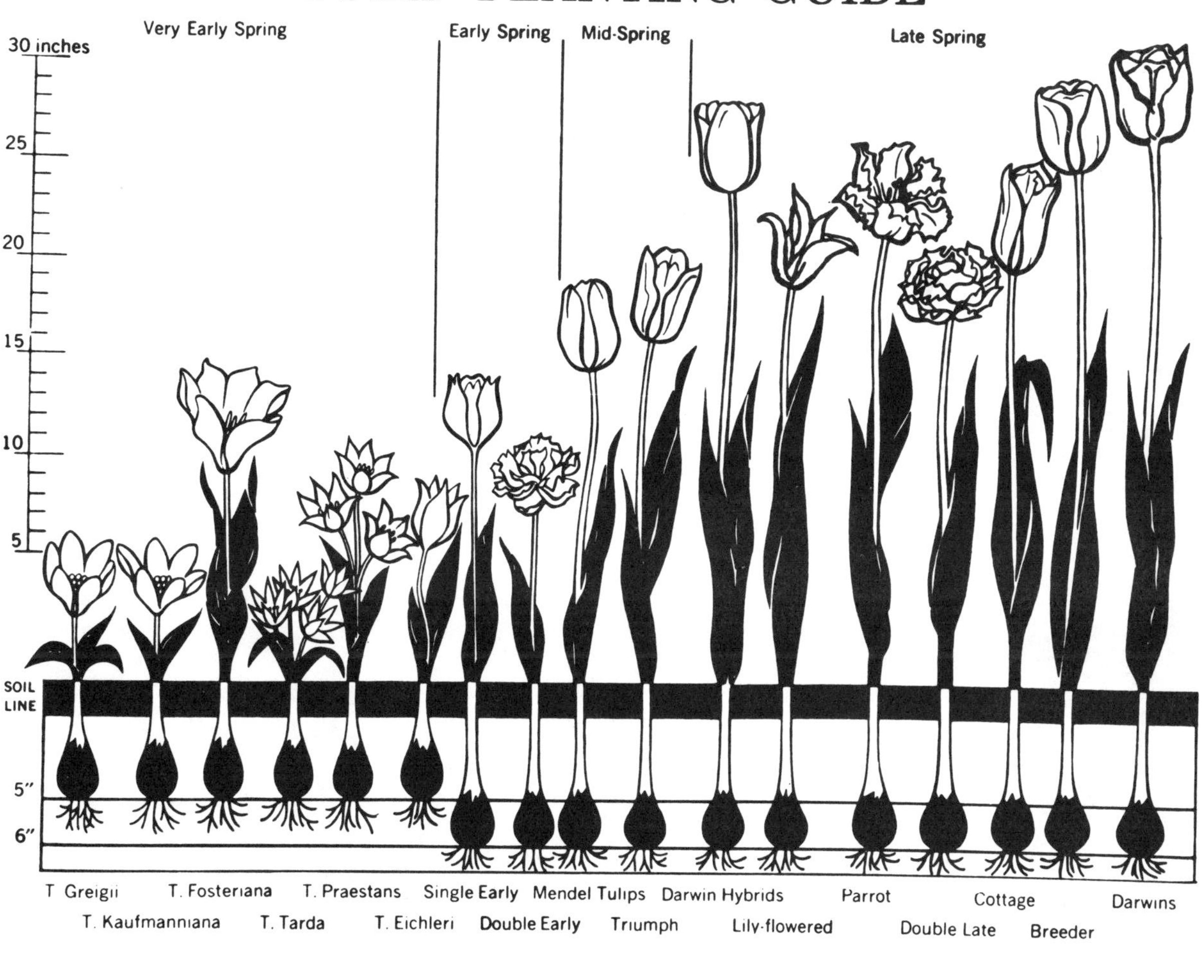

PESTS OF HOUSE PLANTS

GROWING PLANTS INDOORS

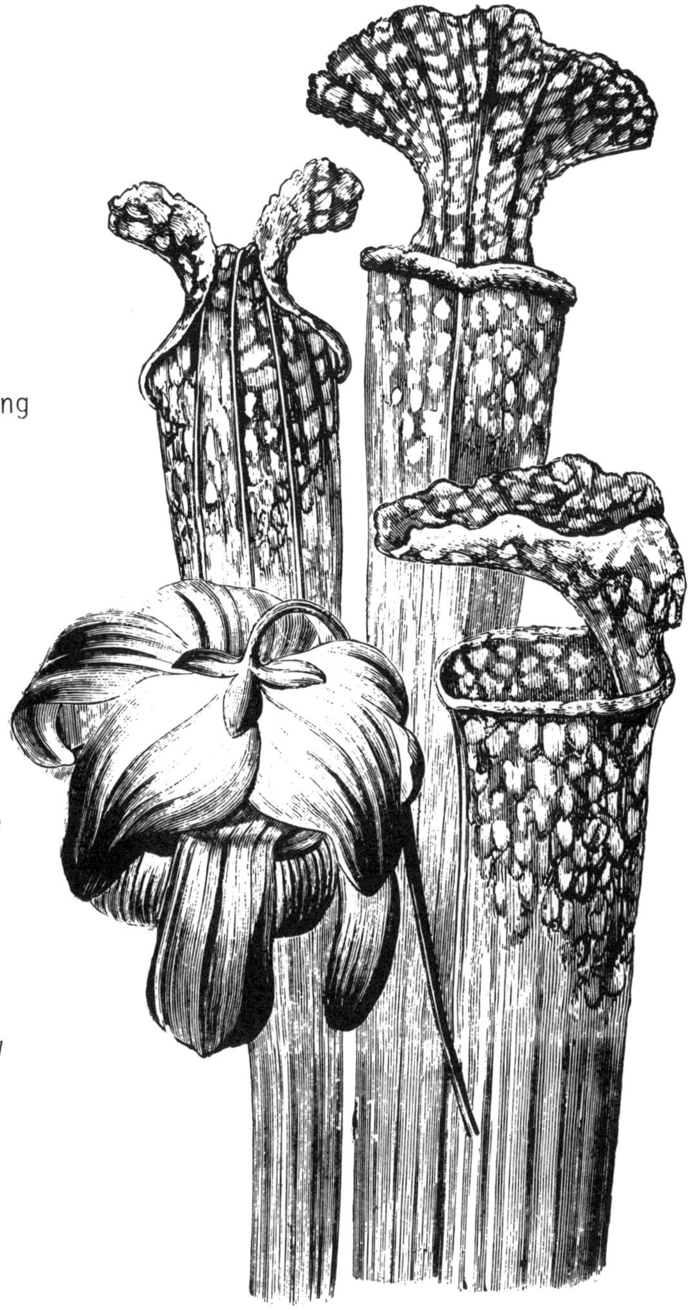

Why do you want indoor plants? Do you want plants that are ornamental and do not require much care, or do you actually want to get involved in the cultivation and care of sub-tropical and tropical rain-forest foliage plants or desert cacti? There is a lot of difference between just collecting and actively growing plants indoors. If you want some plants to put on the table, buy healthy, mature plants that are slow growing and need little care. If you follow a few simple watering and fertilizing procedures your task should be very easy. Many house plant books (The Gardener's Catalogue) tell you what to do if you are to have complete success growing plants indoors.

The following is a list of terms and descriptions of house plant culture that probably will help you determine where to put the plant you have and how to care for it. When you buy a plant from a garden store or plant shop, ask the professional on the staff what the requirements are for that plant. Write them down and follow this list as well.

1. Full sunlight: unbroken sunlight, usually from the South.

2. Subdued daylight: no direct sun, usually from the North.

3. Dim light: any light away from the windows or direct source of light.

4. Indirect sunlight: diffused light as through cutains.

5. Low humidity: 40-50 percent relative humidity. This is normal conditions in a heated or air conditioned room/

6. High humidity: any place that has a high satuation of moisture, such as a greenhouse or hot bed.

7. Medium humidity: relative humidity of about 70 percent.

8. Cool temperature: the temperature on a window in the winter months could range from 40-45 degrees at night, 55-60 on bright days, and 45-50 on cloudy days.

9. Moderate temperatures: normally heated rooms in the winter: 50-55 degrees at night, 62-72 on bright sunny days, and around 60 degrees on cloudy days.

10. Warm temperatures: 65 degrees at night and above 80 degrees during the day.

11. Humus: pure sphagnum moss or osmunda fiber for orchids and certain rain forest plants.

12. Humus soil: a soil mixture of part sand and part humus.

13. Potting mixture: a common potting mixture for most house plants consists of: equal parts of a good sandy loam (basic potting soil) found in garden centers, humus, peatmoss, and any inert mix to lighten the soil and provide good drainage.

14. Growing plants in dim or no light: there are many good books on indoor gardening under lights, so this will just provide a brief outline to help you keep your plants healthy. Prevention is important

a. Most foliage plants can be acclimated to low light.

b. Reduce fertilizer intake of plants.

c. Keep room as cool as possible.

d. Water plants only to prevent wilting.

Watering is an important part of growing plants indoors. Where you have little or no light, start the watering normally and gradually taper it down to where you are watering just enough to prevent wilting. A few leaves may fall off the bottom of the plant. This is normal in the readjustment of the plant to dryer soil conditions. Never let a plant wilt completely. A low temperature will help plants indoors without too much light. Most plants need temperatures ranging from 60 to 75 degrees and low light plants do best at the lower end of the temperature scale.

Begin a fertilizing procedure that will slowly taper the plants down to about one third of their normal intake under ideal lighting conditions. Lighting plants is a very complicated procedure and I would recommend you buying a book (The Gardener's Catalogue) for lighting instructions from some of the biggest manufacturers of the equipment.

INSECTS OF HOUSE PLANTS

ANTS

Ants do little damage or no harm to mature plants. The biggest problem seems to be the tunneling and carrying away of newly planted seeds or very small seedlings. Ants of some species are attracted to aphids, mealybugs, or scale insects, which excrete a honey-dew. The ants feed on this sweet, sticky liquid. To control ants you must control aphids, mealybugs and scale. This can be done in many ways. If you use chemicals, there are many brands and mixes on the commerical market. I do not suggest anything that is poison for indoor use, but most of the brands are the same if you insist. I have never gotten too excited over a few ants, but then I do not get too excited over a few of anything unless it threatens my existence.

APHIDS

There are several species of aphids that do a lot of damage to certain house plants. The common aphids are one six-teenth to one eighth of an inch long and are green, red, pink, or sometimes black. They have pear-shaped bodies with long legs and antennae. All species have winged and wingless types. The wings are held above the body in a roof like position when at rest. Some species look waxy or woolly- this is merely a covering on the bodies.

Aphids cluster on the underside of leaves on foliage plants, or on the buds and new growth of both flowering and foliage plants. There are some aphids that feed on root sections beneath the soil line. Aphids are sucking insects and feed on the juices of the plant. This causes the plant to lose vigor, and causes poor growth, or the whole plant may become stunted. The foliage tends to become distorted and curl-up. All aphids ex-crete a sticky substance called honeydew. This is a sweet material that attracts ants and if left provides a base for sooty mold.

Control of aphids is simple and depends on your preference for chemicals, washing the plants in water under the tap, or using a non-toxic spray such as Cedar-a-flora. You can use a cotton swab dipped in alcohol and just remove the aphids from the foliage by wiping them off.

Cacti are good house plants

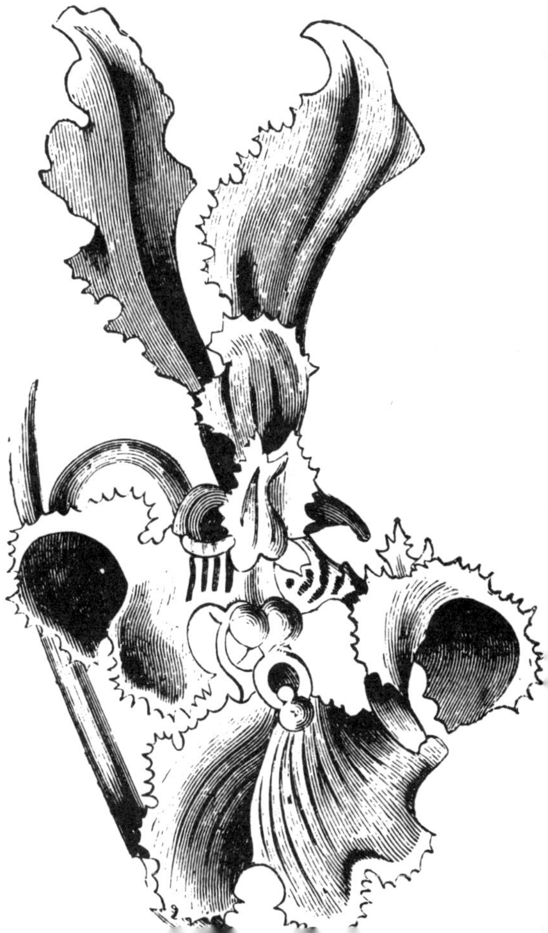

CUTWORMS AND CATERPILLARS

Not too common in a city apartment or an urban setting, but a real problem in the home greenhouse or in a setting where plants are transported inside and outside during the warm and cold months of the year. When hatched you can hardly see the cutworm but they grow to one to one and a half inches long. They are night feeders and hide in the soil or deep in the foliage of the plants during the daylight hours. See the drawing in the chapter on vegetables for a discription of the cutworm and the caterpillar. Some species are solid colors, others are striped and others are mottled. They come in all colors, some are green, brown, or shades of both green and brown. Others are combinations of colors with stripes and a few have tufts of hair on their bodies.

Worms and caterpillars eat foliage. The cutworm cuts off small plants at the soil line or the flowers of larger plants. Control of cutworms or caterpillars in the house or small greenhouse is usually handpicking. If you have large beds indoors or greenhouse benches, you can use a soil insecticide.

CYCLAMEN MITE

Adult mites are too small to be seen with the naked eye. If you can find one and place it under a magnifying glass you will see that the adults are oval, amber, or tan colored, semitransparent and the eggs are white. Mites eat in hidden places on the plant and move from plant to plant only when the foliage touches or they are moved by human means like clothing or hands.

The sucking action of the mites causes the leaves to curl and become twisted. New buds may become deformed and fail to open. The flowers often are streaked with darker color. The foliage and soft stem tissue may become black. A common host is the African Violet. They will develop twisted hairy leaves and may die. Control of mites will be covered at the end of this section on chemical and natural controls of house plants.

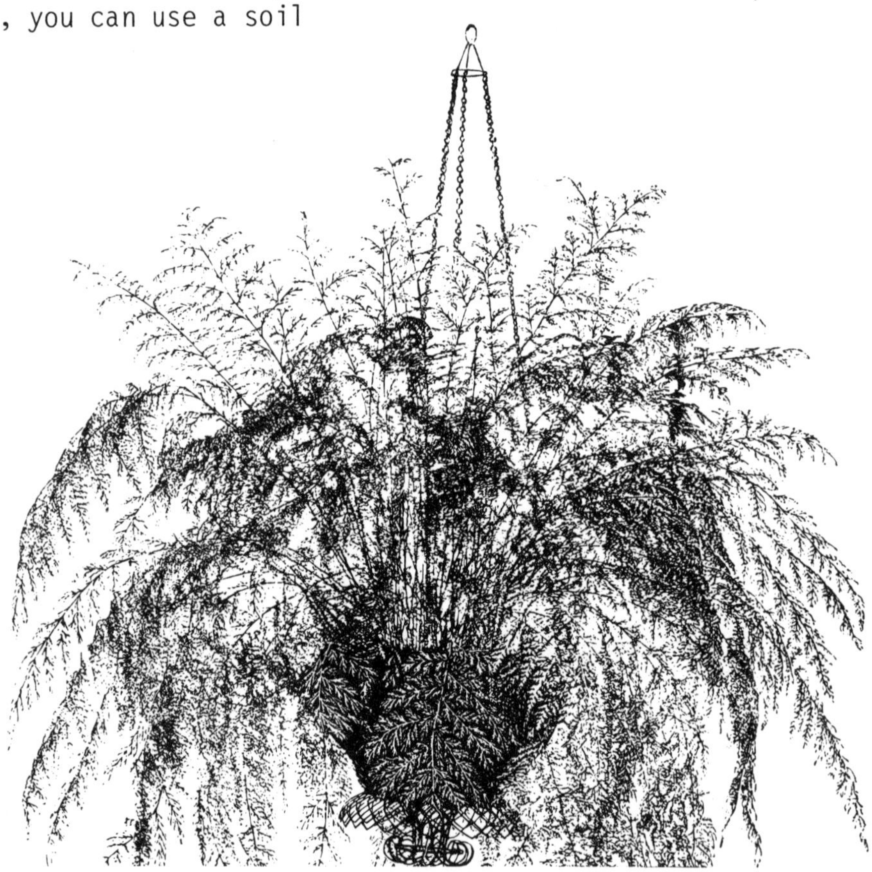

Hanging baskets take more care than plants in trays. You can forget to water them or treat them for insect or disease infestations. Look up once in a while.

FALSE SPIDER MITES

All stages of the false spider mite cycle from the egg to the adult are found on the underside of the leaves along the veins. The adults are flat, oval, dark-red, and too small to be seen with the naked eye. As the infestation becomes severe, the leaf veins tend to become reddish-orange color, and the edges of the leaf may become brittle. The leaves will drop off the plant and, if the plant is totally infested, will die. Look underneath the foliage for signs of mites, aphids, and other insects that feed in this area. (A good rule of thumb is always look underneath the foliage before buying anything in a plant store. Look closely because the pests are small, and if you carry infested plants home and place them beside healthy plants you soon may have a whole tray of mite, aphid, or mealybug infested plants.) You can dip the plants in a mild detergent to wet the foliage and use a chemical dip to control mites. I suggest spraying with Ced-a-flora, a non-toxic indoor insecticide. This product is available in most plant stores. At the end of this section I will list the controls for house plant pests.

FUNGUS GNATS

The adult is a light or dark gray, flylike insect that is one eighth of an inch long. They are attracted to bright windows or bulbs and will swarm near the source of light. The immature gnats are maggots that live in the soil, are whitish in color, and measure about one fourth of an inch. They prefer a soil with a decaying organic matter. If you bring the soil in from the garden (with organic matter or decayed plant parts), the possibility may exist that you might also carry along the eggs of the fungus gnat or other insects. It is impossible to tell if you have the eggs but you can see the maggots.

The maggots of the fungus gnat cause damage to the root systems by tunneling into the soil. They feed on the crowns and roots, and the plants may become discolored and drop foliage. The adults do no damage. But who wants a lot of gnats flying around in the livingroom?

The best control is clean soil and clean pots. Wash the pots with hot water and soap and try to use sterilized soil that you can buy or steam yourself. If you must use outside garden soil for your house plants take all precautions to clean it up. Check it for insects of all kinds. Look in the soil, or run the soil through a strainer or wire mesh of some sort. This is a common practice.

MEALYBUGS

There are several species that do damage to house plants. Mealybugs are softbodied insects that cause the plants to look as if they have been dusted with flour. This appearance is a waxy covering. They are a little over three-sixteenth of an inch long and some species have waxy filaments extended from the rear of their body. You usually find mealybugs near joints where the stem and the leaves come together. They move very slowly and stay out of light just beneath the stems on the leaves. They lay their eggs in clusters enclosed in white, waxy, fuzzy material. For the same reason as the aphid, the mealybug is seen in the company of ants. The ants feed on the honeydew that both insects excrete. The soil mealybug feeds on the roots of African violets and other house plants.

Mealybugs, like aphids, suck the juices out of the plants, stunt the growth, and disfigure the plants. Sooty mold grows

A. Coleus stunted by mealybugs; a. cottony masses surrounding the insects.
B. Healthy coleus plant.
C. Life stages: b. female surrounded by "cotton" and egg; c. small "crawlers" d. crawler somewhat larger; and e., adult male; (all greatly enlarged).

on the excreted honeydew liquid causing still more problems for the plant. If a plant wilts between waterings, it could be ground mealybugs feeding on the root hairs or tender parts of the root system. Control of mealybugs will be covered at the end of this section.

MILLIPEDES

Millipedes are seldom a problem in the house but do cause some damage in the humid home greenhouse. If you have a window greenhouse or an attached small greenhouse on the back porch, keep the area clean of old lumber or flower pots, for this is where the millipedes like to hide. They like organic matter, are active at night, and tend to fold up like a coin if disturbed.

They are about one and a half inches long with many short legs along the sides of their bodies. See drawing for description. They primarily feed on decaying organic matter but will eat tubers, bulbs, and the tender parts of new plants.

Millipede.

Slug.

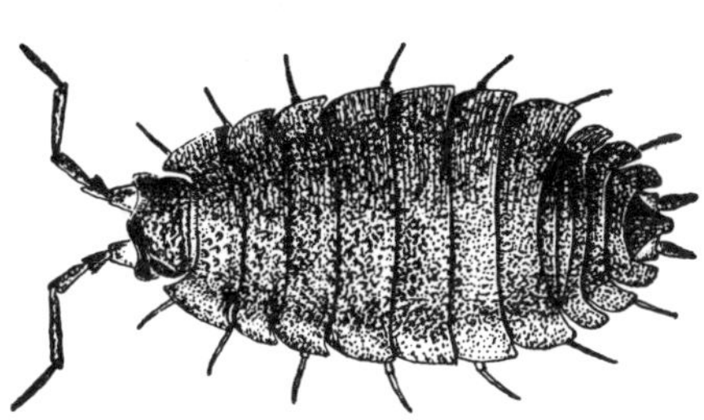

Sowbug, or pillbug.

PSOCIDS

These pale yellowish to white insects have soft, oval bodies and are tiny. Some species have wings and others are wingless. They feed on organic matter and become a nuisance for they tend to gather in large numbers. They do not feed on living plants but are unsightly. They like damp spots like millipedes do, and will live in undisturbed places in the greenhouse or potting shed. You just have to clean out the collected junk once in a while.

SCALES

Scale are sucking insects, and several species invade and live on house plants in the home and greenhouse. Most species are one-sixteenth to one-eighth of an inch in diameter but a few species are a little larger. They are hard shelled insects that seldom move once they start feeding. The colors range from black to white, but brown scale is the most common in the home greenhouse and window boxes that grow house plants.

The eggs of most species are white and are laid in a white sack secreted from the scale. This egg cluster is often mistaken for mealybugs. The scale attacks all parts of the plant-stems, foliage, and new shoots. Scale (like aphids and mealybugs) attracts ants for the honeydew. As do the other sucking insects it also provides a base for sooty mold. The dead scale appears dry and chaffy. Controls will be covered at the end of this section.

SLUGS AND SNAILS

See drawings of both slugs and snails. They are not a problem in the house but can cause trouble in the potting shed or home greenhouse. They are slow moving and measure from one quarter of an inch to four inches long. The snails carry a shell and range in color from black to light brown.

Both slugs and snails eat holes in foliage and flowers. They leave a trail as they crawl about the walks or paths of the greenhouse or yard. If you see them in the house, you are in real trouble. (I suggest closing the doors.) There is a commercially prepared bait for slugs sold in all nurseries or plant shops.

SOWBUGS AND PILLBUGS

See the drawing for description. Both sowbugs and pillbugs have segmented shell-like bodies, are one quarter of an inch long to one half an inch, and are gray to brown. If you are growing seedlings in a frame or starting seeds in the shed where there are hiding places or damp organic matter, sowbugs and pillbugs may feed on the new plants. Eliminate all damp hiding places that might serve as a resting or hiding place for the bugs.

SPIDER MITES (RED SPIDER)

Spider mites or red spiders as they are called, come in colors from green to yellow to red and are tiny and oval. They like dry warm conditions. They are usually found clustered on the underside of the leaves - a silky web indicates a really heavy infestation. The webs are strung from leaf to leaf and the mites crawl from plant to plant.

Spider mites are sucking insects (these are not really insects but here we will call them insects) and suck the sap from the stems and foliage. The first signs of damage, after the webbing, is whitish or yellowish spotted areas on the leaves. If the infestation gets out of hand the leaves may become completely yellow and drop off the plant. The flowers of the plant may look faded. This pest may kill the whole plant.

The first step is to break up the webbing with a sprayer. This will dislodge the spider mites and leave them in isolated spots. Dip the plants or hand clean very tender parts of the plant. For complete instructions for controls see the end of this section.

SPRINGTAILS

Although, these pests do not cause damage to growing plants they are a nuisance for they usually are found in large numbers. They feed on organic matter in potting rooms or the greenhouse. Not a problem in the house. Springtails range in size from microscopic to about one fifth of an inch long. They are whitish to black and sometimes tinted blue or purple. They jump. As with other pests of this nature, keep the potting shed free of loose organic matter that might collect damp conditions.

THRIPS

These are slender, barely visible pests that are tan, brown, blackish-brown, or black with lighter markings. The young are white to yellow or orange and some species carry around droplets of black excrement on their backs. The young are less active than the adults but will fly about (really jumping) if disturbed. Both adults and the young thrips suck the juices of house plants. The injury can be detected by silvery streaks on the foliage, little black dots on the leaves which are the excrement of the thrip, the foliage may blotch or droop, and the flowers may become distorted or streaked.

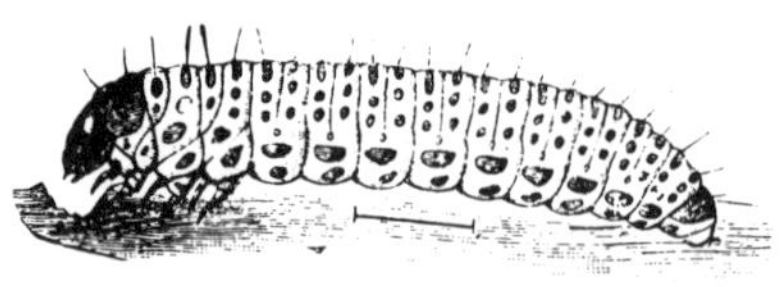

WHITEFLIES

Whitefly is a troublesome pest if found indoors. Most commercial greenhouses just close the ventilators of the house and completely fumigate the place, but that is impossible to do in the house. I remember reading about a New York State **botanist,** who was working as an extension service agent in a weekend answering service to help home gardeners with their problems. He suggested using a vacuum cleaner and sucking up the white cloud of flies into the waiting bag. Although he was just kidding around, in reality, this is not a bad idea. I suspect he was just a little tired of answering questions about flies, twigs, burns, spots and assorted mishaps.

Whiteflies are about one sixteenth of an inch long and have white wings. When you move a plant the flies swarm around in the air like snowflakes in a windstorm. The scale-like young are green to yellow or white, oval in outline, and flat on the top. All stages of this pest are usually found on the underside of the foliage.

Damage is caused by the sucking of the juices of the leaves, stems and tender new shoots. The infested parts become pale and turn yellow. Like all sucking insects, the surface of the plant that has been affected is covered with a sticky substance called honeydew and this serves as a base for sooty mold. Control of whitefly is tough, and I suggest if you only have one plant that has a heavy infestation, to remove it from the surrounding plants. For further controls look at the end of this section.

CONTROL OF HOUSE PLANT PESTS

You can control most of the pests of house plants with sprays and dips. The exceptions are slugs and snails that are controlled either by hand or poison bait. All the insecticides that are toxic should be applied as a dip. I would not recommend spraying any chemical in the house. I would take all my plants outside in the warmer months to clean, dip, or spray them. During the colder months, dip them with an insecticide solution. Read the guide for mixing sprays in the insecticide section. You can prepare all your own dips at home. The push-button sprays are not recommended at all.

The basic ingredients of a spray or dip are:

1. The insecticide in the form of a wettable powder or an emulsifiable concentrate.

2. Water. (The solutions made with emulsifiable concentrates leave less residue on the foliage and tend to wet the plants better than the wettable powders. The bad news is that the emulsifiable concentrates will burn the foliage if not prepared exactly in the right way. Follow the table in the insecticide section.

3. If you buy a product with a mixing instruction key, and this differs from the mixing guide, use the solution with more water.

4. Add the concentrate to the water and stir until they are completely mixed. Make only what you need. Try not to store chemicals.

5. If you buy a commercial procuct with two or more chemicals included, follow the manufacturer's instructions. Write the manufacturer for any additional information you might desire.

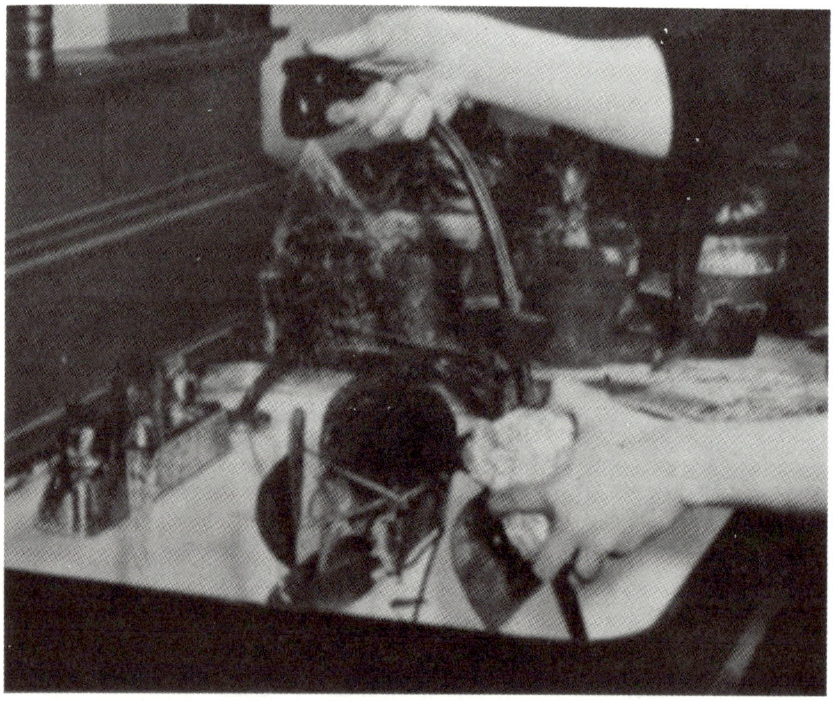

Spraying a broad-leaved plant with lukewarm water- one way to remove aphids, mealybugs, thrips, and spider mites.

APPLICATION OF SPRAYS AND DIPS

Many plants are hard to wet, so add a teaspoon of detergent (not soap) per gallon of mixture. The detergent improves the wetting power of the mix. While applying the spray or dip, continue to mix or shake the solution as the active ingredients tend to fall to the bottom of the container. Spray plants until they are wet. Use a sprayer that is non-corrosive, like stainless steel. After using a sprayer with insecticides wash it out - you don't want chemicals in the sprayer when you use it as a mister.

HAND ATOMIZERS

An adjustable nozzle that delivers a continuous spray is the best type. It gives a fine mist and covers more of the surface of the plant. The old-fashioned pump types are still around but they throw more heavy water and less mist. An investment of a good pressurized hand atomizer will not only serve your insecticide spraying needs but will serve as a mister. You will need to mist rain forest plants in an apartment or house, where the humidity level is much lower than outside conditions. You may want to buy a humidifer if you have a lot of foliage plants.

DIPS

Dipping plants is the best way to get rid of most house plant pests. Just put the solution in a pail and dip the plant in it. I have a cardboard circle with a slit half way through the center that I hold over the soil as I dip the plants in the pail upside down. Wear rubber gloves for your own protection when dipping or spraying.

OTHER CONTROLS

1. Selective buying (inspect plants for health).

2. Regular water with a mild detergent (just wash them off the plant).

3. Hand pick the bugs.

STRAWBERRY JAR PLANTED WITH HENS AND CHICKS.

Some causes of unhealthy appearance in house plants

Symptom	Possible cause
General defoliation	Sudden change in temperature. Transplanting shock. Sudden change in lighting intensity—moved from strong sunlight to a dark location. Overwatering.
Leaves drop, shoots remain dwarfed and branch repeatedly. New leaves are small.	Manufactured-gas injury (unburned cooking gas in the atmosphere).
Browning of leaf tips	Improper watering. Exposure to cold drafts. Insect attack. Excess fertilizer.
Spotted foliage	Overwatering. Burning from direct sunlight.
Loss of normal foliage color	Overwatering. Lack of fertilizer. Insect attack.

PESTS OF LAWNS

LAWN DISEASES

Most of our lawns are grown under what might be considered artifical conditions so suffer disease attacks more readily. A field grass, well established and growing wild, does not suffer from disease as much as a bluegrass growing in new topsoil that is only six inches deep. Many of the "new grasses" are very beautiful and fast growers but seem to be more susceptible to plant diseases and insect attacks. Of course the best method of controlling lawn diseases is to maintain a properly managed lawn, feeding it regularly, and applying the right chemical (if that is your way) in season to prevent any problem from taking hold and spreading. The other solution is just let it all grow wild and go fishing. But a nice lawn needs basic requirements.

1. Buy grasses suited for your climate (see grass map).

2. Proper mowing and raking.

3. Yearly feeding at the right time.

4. Proper watering at the right time of day during the hot season.

If you follow these guidelines you will have little or no problems. But if you do spot something that looks out of place and does not match the color or texture of the established grass in your lawn, you might suspect some causal problems which some gardeners mistake for diseases: undesirable grasses for your climate; chemical (herbicide) and fertilizer burns; dog urine; septic tank damage; dry compacted soil; localized dry or wet spots; poor drainage; too much shade for direct sunlight varieties; too much sun for shady-spot varieties; and finally the number one mechanical problem of the lawn, thatch (grass cuttings that are not picked up accumulate just above the soil line, holding all the moisture and causing the root sections of the grass to turn up).

FUNGUS DISEASES OF LAWNS

By far the worst problems of lawns are fungus diseases. Other causes are not serious enough to spend time on, and if you cannot recognize the trouble from this book, I suggest you consult your nearest professional turf-person or extension service agent. Lawn fungi appear on grasses in the form of masses of microscopic threads or filaments called mycelium that often have the appearance of cobwebs. The spore is the fruiting body that reproduces the disease. These fungi get all their food from the host plants (grass). There are several main kinds:

LEAFSPOT

BROWN SPOT

RUST

DOLLAR SPOT

SNOW MOLD

STRIPE SMUT

FADING OUT

Slime mold and mushrooms do some damage in lawns but they do not attack grasses.

BROWN PATCH

This fungus attacks all forms of grasses but is most common in bluegrass, fescue, centipedegrass, St. Augustinegrass, and ryegrasses. This is a problem throughout the United States but is more common in the regions where hot, humid weather dominates the summer months. This problem may occur just after a heavy feeding of the lawn with a nitrogen fertilizer. The nitrogen produces a fast green growth which is the breeding ground for brown patch. If the disease is present in the area it can spread by the mower or even on your shoes as you play or walk on the grass. The disease spreads by fungus threads or mycelium.

Symptoms:
Brownish patches develop on the lawn. They may be from a few inches to a few feet. In bentgrass a narrow, dark colored ring surrounds the infested area. As long as the rhizomes and the crowns are not killed the plant will recover. A little discoloration of the leaves is a cause of concern but may not kill the plant. If the weather conditions remain hot and humid, the disease will spread and kill the crown and the whole plant. When the weather becomes dryer and somewhat cooler, the symptoms may totally disappear.

If you see filmy, white tufts on the grass in the morning during the time of the day when the grass is still wet with dew, this probably is fungus threads of brown patch. All fungus needs moisture to reproduce and spread so it would not be wise to water the lawn if you think a fungus is growing. Let the lawn dry out and cut the bad sections. Bag and remove the bad grass.

DOLLAR SPOT

This is also known as small brown patch because the appearance is almost identical. It bothers most varieties of grasses but particiularly bentgrass. You can find this disease in most regions of the United States and Canada, but generally prefers humid northern areas. The attack begins in the early summer when the weather is cool and humid, say around May and early June. It lets up during the hot summer months but will begin again in the fall periods when the weather is again cool and humid. The plants are more susceptible if the soil has a nitrogen deficiency. Keep a good amount of available nitrogen in the soil and you may help to control this disease.

Symptoms:
The spots are round and about the size of a silver dollar. First the grass is dark and appears somewhat water-soaked, then it turns brownish color, and finally white and bleached out. If the disease gets out of hand the spots may merge and form large irregular areas. If you catch this disease in the early stages you may only have a slight problem. Use the chemical chart in this section for control.

FADING OUT

This is a fungus attack that bothers bluegrasses, bentgrasses, and fescues, and is common throughout the United States and Canada. It is caused by a complex of Helminthosporium and Curvularia. Most of the serious damage is done during hot, humid weather from May until frost.

The grass that has been affected by this disease appears yellowish with spotting of green. It may look as if the grass is suffering from an iron deficiency or lack of fertilizer. When this disease reaches the "out of control" stage it leaves patches of reddish-brown areas sometimes two to three inches in diameter, which blend together and form large dead spots in the lawn. The same practice used for leafspot can be used for this disease.

HELMINTHOSPORIUM LEAFSPOT AND

FOOT ROT

Kentucky bluegrass suffers as a result of this disease that gets its name from the helminthosporium fungi. This is a common disease found in most parts of the country. During cool moist weather this fungus develops at a very fast pace.

The fungus that causes leafspot in Kentucky bluegrass also causes a condition as going out or melting out. The disease is more likely to attack if you have a pure stand of bluegrass - a mixture of fescue or redtop may help. Many species of grasses are naturally resistant to this disease so buy or prepare a mix that will prevent a bad outbreak.

Symptoms: A reddish-brown to black spot appears on the leaves and the stems of the grass plant. Leaves shrivel up and the stems, roots, and rhizomes discolor and die. Leaf-spot on other varieties of grasses produced by the same fungi have the same appearance as the symptoms on Kentucky bluegrass. The areas left open by the rotting and dying of the rhizomes are invaded by weeds and other undesirable grasses such as crabgrass. Control: Grow disease resistant varieties such as Fyling, Windsor, Merion, and Pennstar. While these varieties may have some leaf-spotting, they seldom are totally killed during the rotting stages of the disease as are less resistant strains.

Mow the grass in an upright position to about one and a half to two inches tall. Always pick up the clipping. Do not overfeed with a heavy nitrogen fertilizer in the spring. Nitrogen is the first number on the fertilizer bag label, and anything over ten should be avoided. If you use chemicals you can use a fungicide. See Fungicide Chart at the end of this section for proper use.

SMUT GRASS

DROP SEED-NIMBLE WILL

SCHRADER'S GRASS

MUSHROOMS AND FAIRY RINGS

The real problem with mushrooms in the home lawn is the poisonous nature of many varieties. Children and animals pick and eat them and end up in the emergency ward of the local hospital. It is very natural for mushrooms to develop but you can help control them by following a few rules: Any buried organic material like stumps or old fill should be dug up and removed. You can then reseed the spot. A well drained lawn will help keep the mushroom population down.

Fairy rings are circles or half circles of dark green grass surrounding areas of light colored, dead grass. During the spring of the year the mushrooms develop in a circle outlining the fairy ring. If you do not check the fungus growth, the circle will enlarge each year.

This fungus grows in an outward motion from the starting point, and can spread from ten to thirty inches a year depending on local conditions. This fungus is found several inches below the soil surface and it forms a dense layer of mycelial threads that break down the organic matter at the outer edge of the ring. See lawn chart for proper controls.

PYTHIUM DISEASES

Cottony blight and grease spot are the two most common diseases caused by the fungus pythium. Cottony blight only occurs in the South on ryegrass but grease spot occurs throughout the country on many strains of grasses. If the temperature is above 70 degrees with a high humidity level, and if the soil is poor and improperly drained, this disease may appear more readily. Even on a new lawn you might find an outbreak of the fungus if the weather conditions and the soil conditions are favorable.

TALL REDTOP

Symptoms: The areas infested with pythium diseases may be from a few inches to a few feet long. You can see the spots on the lawn in the early morning. They are from one to two inches in diameter surrounded by blackened grass blades that are mixed with the fungus threads. After the disease has become established, the grass appears slimy and matted- the grass will lie flat (not like brown patch that has a tendency to remain upright after the grass has died). The dark blades of grass soon wither and become brownish-red. The plants die within a few days after the disease has gotten a firm footing. You will have to reseed the brown areas for new grass does not grow back in the diseased spots.

Control: The best advice is not to over water lawns. Do not let the sprinkler remain on for a long period of time and completely saturate the soil and plants, as this will cause the condition that attracts this fungus. Plant new seed in the fall when the weather is cool.

RUST

Like brown patch, rust favors a moist grass to reproduce. It usually is found from late summer until frost. Grasses more susceptible to rust are: Merion blugrass (wherever grown), and Zoysias (except Meyer and Emerald).

Symptoms: A yellowish-orange dust (the spores) appear on the leaves and stems. If you rub a white cloth across the lawn you think is infested, the cloth will pick up the above colors.
Control: If you have a pure Merion bluegrass lawn you can expect more trouble than a grass mix. Good mixes to avoid a lot of trouble are:

50 percent Merion and 50 percent red fescue.
50 percent Merion and 25 percent each of Kentucky bluegrass and red fescue.

If you use chemicals to control diseases check the chart at the end of this section for proper control of rust.

SLIME MOLDS

This mold does not feed off healthy grass but on dead organic matter, such as old grass cuttings. This group of fungus which range from bluish-gray to yellow black, is unsightly on a well-groomed lawn. The only damage they do is to shade the blades of grass and slightly discolor the stems and blades. They only occur during wet weather and usually disappear when it becomes hot and dry. They are easily swept away with a rake or lawn broom. If you have a long wet spell, and want to apply a fungicide, check the guide for selecting fungicides at the end of this section.

SNOW MOLD AND FUSARIUM PATCH

Sometimes called winter scald, snow mold hits the bentgrasses harder than other strains. It is caused by several fungi. When grass is covered by snow for long periods of time it is very hard to control snow mold. If the grass is green and growing it is almost impossible. Fusarium patch, also called pink snow mold, will not tolerate temperatures above the low sixties but will spring up in a moist, low temperature lawn. Poor drainage during low temperature times will encourage both of these fungi.

Symptoms: A white, soft growth appears first on the tops of the grass with snow mold. As the disease progresses, the blades turn brown and stick together. Diseased areas range from a few inches to several feet in dimension. They have a discolored dirty look of grayish-white turning pink. Fusarium patch appears as irregular patches of pale yellow, later turning a light gray. Many times the affected areas have a pinkish color around the edges.
Control: Good turf management in the fall is the key. Never apply a heavy fertilizer in the late fall. This may encourage growth activity under the snow. Before you get the first frost or snow, clean and rake the lawn. Never allow any matted areas to remain under the snow. Apply lime if a soil test or local conditions indicate need. You can test yourself or you can get the testing done by your extension service. See the chemical chart at the end of this section for control of molds.

STRIPE SMUT

This disease attacks many strains of cultivated grasses but is most common in Merion bluegrass. It is found in the northern regions of the country. The spores of this fungus germinate in the soil and produce threads that invade young grasses and tillers. This fungus is not like other fungi as it grows in a very systematic way and will stay with the plant until it dies.
Symptoms: Narrow black and gray stripes develop lengthwise on the blades. The stripes may be continuous or discontinuous. The gray stripes are unruptured smut called sori and the black stripes result when the smut sori rupture and liberate the black masses of spores. After the smut ruptures the grasses wither, turn downward, and die.
Control: Grow disease resistant varieties like Park, Flyking, and Pennstar. If you have to grow Merion bluegrass, mix it with common Kentucky bluegrass. See the guide for selecting fungicides at the end of this section.

LAWN INSECTS

ANTS

The problem with ants is the general habit of this insect to build nests in grass and mound up the soil, which often will destroy grass by smothering it. Sometimes the nesting in the soil will destroy root sections of the grass. In the West the winged harvester ants have a tendency to swarm and may do a lot of damage to newly planted lawns. The fire and harvester ants may bite small children and pets.

ASIATIC GARDEN BEETLE

This beetle measures about one quarter of an inch, is chestnut brown, and has a velvety color texture. The underside of the beetle is covered with yellow hairs. It flies only at night. They are found primarily on the Atlantic Coast and Eastern New York State, and can be seen anytime from mid-July to mid-August.

BEES, WILD

Not a major pest, but some bees do dig in the soil and build mounds.

BILLBUGS

The grubs are small and white and have brown or yellow heads. They feed on the roots. The adult billbug is about one half an inch long and has a long snout with a pair of jaws for chewing. They may be yellow to reddish or very black (see photograph). The beetles also do damage by burrowing in the stems and feeding on the blades. There are several species of billbugs that do damage: Sphenophorus phoeniciensis attacks bermudagrass in the Southwest and California; S. cicatristriatus found in the state of Washington. The hunting billbug, which has caused damage to zoysia grass on the Eastern Seaboard, in California, and Hawaii.

CICADA-KILLER WASP

This wasp builds nests in the soil. It is about one and a half inches long with yellow and black markings. The female paralyzes a cicada (see periodical cicada) by stinging it, and then placing it in her nest and lays eggs on it. When the eggs hatch, the larvae feeds on the cicada. Most damage is done in the Eastern part of the country. They appear in late June and can be seen until late August. They also sting people if molested.

EARTHWORMS

The usefulness of this insect outweighs all mounds and castings that might be found on your lawn. If you fish for bass in the early summer or just like to sit on the bank with a bobber, let the worms work.

EUROPEAN CHAFER

This light brown beetle is about one half an inch long. It swarms at night in trees, making a buzzing sound. Only found in parts of New York State and Southern New England.

GREEN JUNE BEETLE

This is a big beetle measuring nearly one inch long. The body is almost flat and has a bronze-to-yellow edge. This beetle feeds on a variety of plants, from small ornamental trees to herbacious garden flowers. The female will lay her eggs in old grass clippings or in the soil. (Just another reason to rake the lawn after mowing). This pest is found primarily in the Southern regions of the United States but has been seen as far north as Southern Illinois. They are most active in the summer months and produce one generation a year. The grubs feed on old organic matter and their burrows may uproot young grass seedlings. After a heavy rain this grub may emerge out of the soil and feed on the surface. They have a habit of crawling on their backs.

GRUBS

These larvae of may kinds of beetles, are usually white or gray with a darker head, and eat the root sections of most of our native grasses. They lie in a curled position. The eggs are laid in the soil by the adult beetle. Depending on the species, many larva spend from ten months to two years about two inches deep in the soil. In very cold weather they go deeper into the soil. The grubs feed on the roots of the grasses about one inch below the surface. Many times areas that have infestations of grubs will be torn up by moles, skunks, and birds feeding on the grubs.

The white grub seems to be the most wide spread and causes the most damage in our lawns. The adult May beetle (see photograph) is brown or blackish-brown. There are over 200 kinds of beetles that fall into this wide category. (They have different names in different parts of the country--some are called June beetles). All produce the white grub. The photograph shows the characteristic curled position of the grub in the soil. In different parts of the country the beetle appears at different times:

South: early April

Southwest: July and August

North Central: May to mid-July

Northeast: Not common but same as North Central

JAPANESE BEETLE

This beetle is just over one half an inch long (see photograph) and is found in all parts of the United States and Canada. It has a shiny metallic shell with green coloring. The wings are copper colored and it has six small patches of white hairs along each side of its body. The adult beetle does a great deal of damage to over two hundred varieties of garden plants. The beetle appears in the Eastern part of the country in early June and in the South as early as May. It seems to be uncommon on the West Coast but has been seen there. They feed until frost in the Northern States. (See the longer discussion of the Japanese beetle under sweet corn in the vegetable section).

MASKED CHAFERS

This half inch long beetle lives in the soil during the day and emerges at night. It has a light to dark brown color. If the weather is warm and humid they become very active. The masked chafer is found on the Eastern Seaboard from Connecticut to Alabama and West to California. In the Southeastern part of the country and in Texas they appear in mid-June and are active for about two months.

The grub of this beetle has irregularly shaped spines on the underside of the last body segment. They are called annual white grubs because they complete their life cycle in one year. (Many of the other white grubs take many years to complete the transition from grub to adult beetle).

VANILLA GRASS

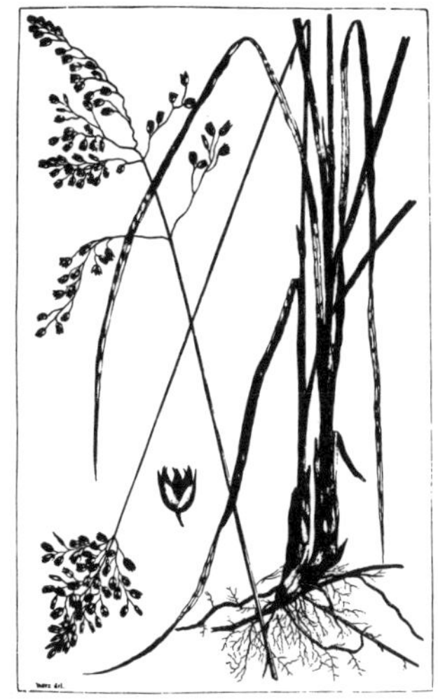

ORIENTAL BEETLE

Only found in the Northeast. They appear in late June, prefer shady places in the lawn, and usually only feed on short grasses.

PERIODICAL CICADA

The nymphs leave small holes in the lawn, especially under larger plants as they emerge from the soil. If you hear the song of the cicada in the spring, a brood is about to be born.

ROSE CHAFER

A yellowish brown beetle about one half an inch long, with long spiny legs. It feeds on a variety of plant forms. This beetle likes a shady, sandy loam soil. Found in the Eastern States and Colorado to Texas. These grubs are not as harmful as the grub of the Japanese beetle but they do some harm.

WILD FESCUE

Damage to grass by Japanese beetle.

Japanese Beetle

Billbug

White grubs.

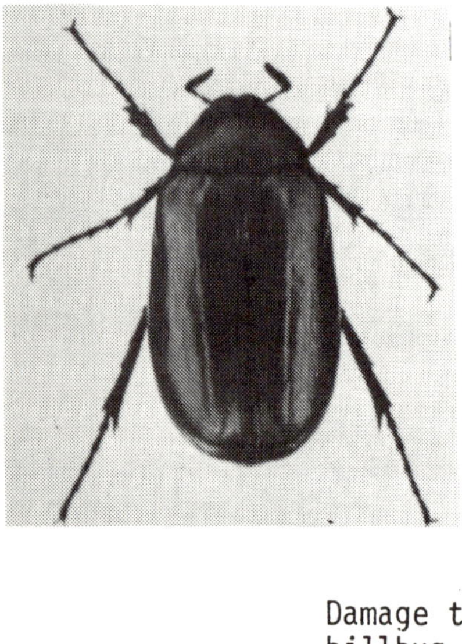

May Beetle

Damage to grass caused by billbug grubs. Arrow points to grub.

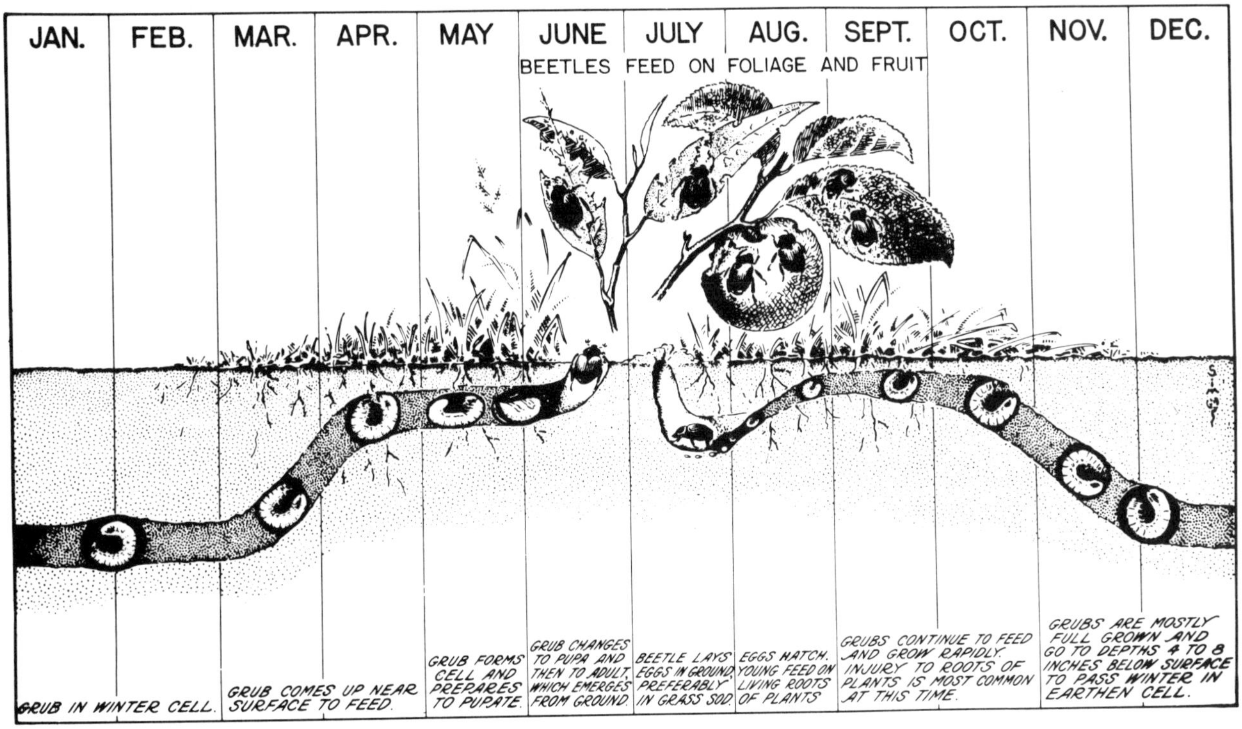

Seasonal life cycle of the Japanese beetle in the vicinity of Philadelphia.

REGIONS OF GRASS ADAPTATIONS

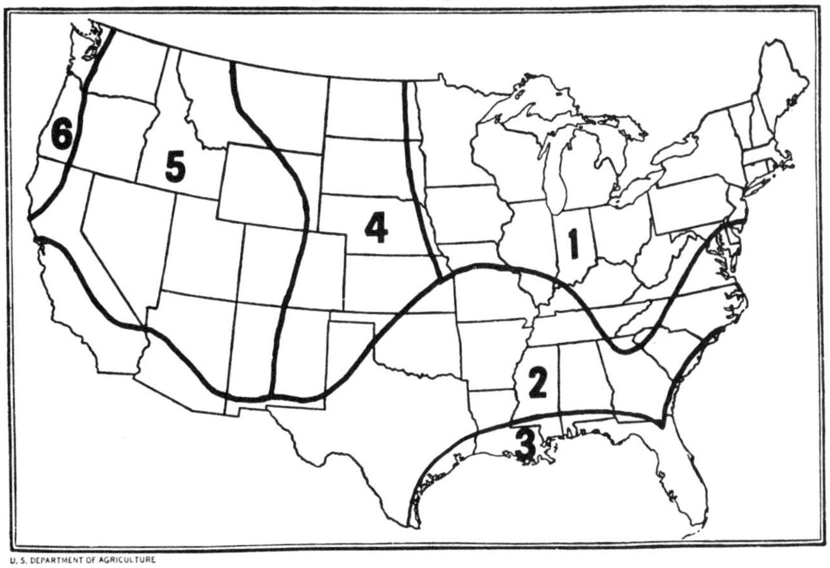

Climatic regions of the U.S. in which the following grasses are suitable for lawns: Region 1. Common Kentucky bluegrass, Merion Kentucky bluegrass, red fescue, and Colonial bentgrass. Tall fescue, bermudagrass, and zoysiagrass in southern portion of the region. Region 2. Bermudagrass and zoysiagrass. Centipedegrass, carpetgrass, and St. Augustinegrass in southern portion of the region with tall fescue and Kentucky bluegrass in some northern areas. Region 3. St. Augustinegrass, bermudagrass, zoysiagrass, carpetgrass, and bahiagrass. Region 4. Nonirrigated areas: Crested wheatgrass, buffalograss, and blue gramagrass. Irrigated areas: Kentucky bluegrass and red fescue. Region 5. Nonirrigated areas: Crested wheatgrass. Irrigated areas: Kentucky bluegrass and red fescue. Region 6. Colonial bentgrass and Kentucky bluegrass.

Damage to grass by Japanese beetle grubs

Rhodesgrass scale

White grubs

Ground pearls

Sod webworm adult

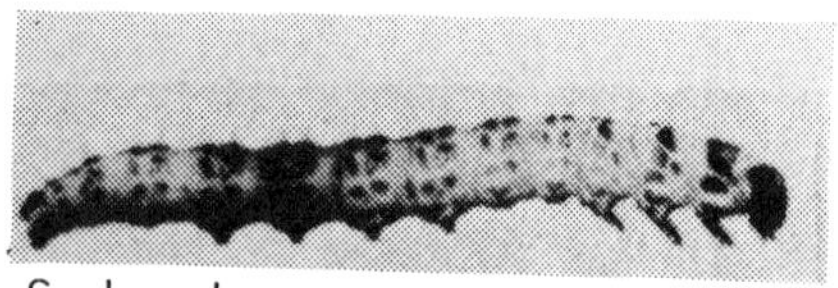

Sod webworm

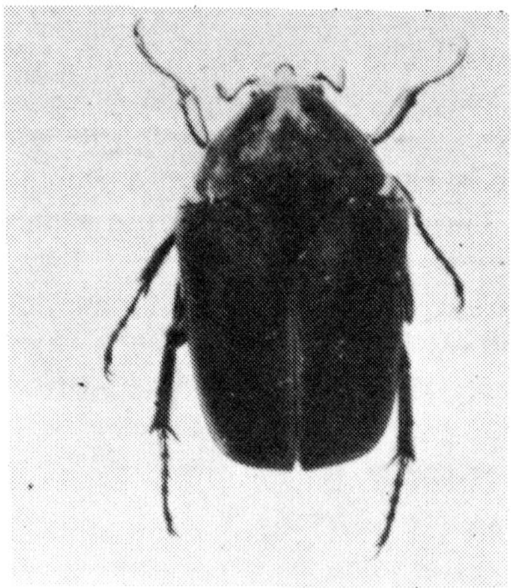

Green June beetle

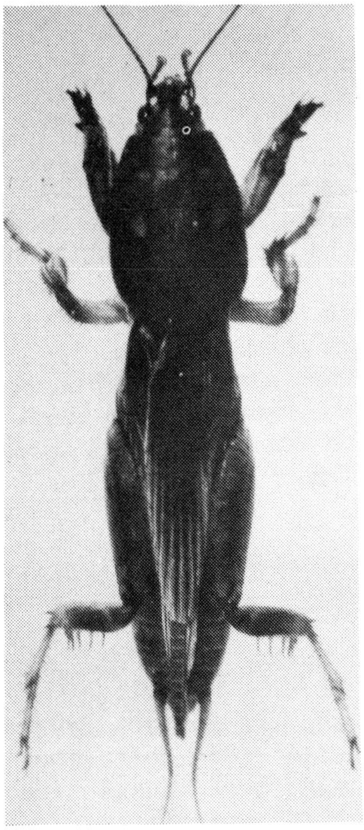

Mole cricket

WIREWORMS

These are the larvae of click beetles and are one half an inch long to one and a half inches. They come in two colors: white or yellowish; or dark brown and very hard. They bore into the underground part of the stems and feed on the roots. The adult beetle is half an inch long, brownish-gray (sometimes almost black). Their bodies taper at each end, and the shell is hard. When they fall on their back they flip the middle part of the body against the surface, throw themselves up in the air and make a clicking sound.

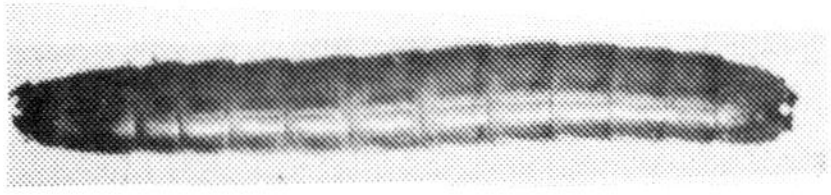

Wireworm

MOLE CRICKETS

Mole crickets feed on the roots of grass. They are light brown in color and have green markings. They are about one and a half inches long and have short, stout forelegs, shovellike feet, and large beady eyes. The burrows made by this insect may uproot young grass and cause a lot of damage in a newly-seeded lawn. Found in the South, primarily in the Gulf Coast States and Texas.

Effluent from the septic-tank sewage disposal system in this backyard is rising to the surface. The soil is poorly drained; septic-tank systems in such soil will not function during wet weather.

ARMYWORMS

Usually about one and a half inches long and greenish with black stripes down each side and center. The adult is a moth with wings about one and a half inches and brownish-gray bodies. The armyworm and the fall armyworm eat a lot of grass in a short time. They are very common and their feeding habits leave circular bare spots in the lawn. See drawing for discription.

CUTWORMS

These are dull brown, gray, or black caterpillars about two inches long. Some have spots on them, others are striped. These caterpillars are the larvae of the night flying brown moth. They feed at night and hide in the soil during the day. They do most damage to the seedlings of bentgrass, bermudagrass, and ryegrass.

FIERY SKIPPER

The larvae of the fiery skipper feed on the blades of most common grasses and leave small round, bare spots. The adults are small yellowish-brown butterflies.

FRIT FLIES

This fly is very small. It lays its eggs on the grasses and then the maggots bore into the stems. It seems to be found in parts of New England. Not a major pest.

GRASSHOPPERS

Not troublesome to a well-kept lawn. They usually feed on agricultural grains and breed in grass wastelands.

LEAF BUG

This gray and white insect feeds on lawn grasses but only causes slight damage. Not a major pest.

MEADOW FESCUE

LUCERNE MOTH

Clover and legumes are favorites of the larva of this moth (they have been known to eat grasses but not usually). The adult is a brownish-gray moth with two pairs of dark spots on each forewing. They have been recorded in California and about thirty other States.

SOD WEBWORMS

This is a light brown worm about three quarters of an inch long with hair completely covering its back. Sod webworms are the larvae of the lawn moth, which is a small whitish-gray insect (sometimes called millers) that fold their wings closely about their bodies when resting. They usually hide during the daylight hours in low growing shrubs. In the early evening they fly over the grass and the female scatters her eggs.

The sod webworm (see illustration) works only at night. They live in protecting silken webs that they spin about their bodies. They begin to feed as soon as they hatch. As they progress in size they begin the building of burrows just beneath the surface of the soil. They reinforce the tunnels with cuttings of grass and small particles of soil, then they line the tunnels with a silken matter - this is home. The sod webworm cuts off blades of grass and feeds on the crowns at soil level and the roots. They over-winter in the soil, in the silken webs.

Sod webworms prefer new grasses and do a lot of damage in newly seeded lawns. They attack bentgrass, bluegrass, and fescue. The first sign of sod webworm damage is irregular brown patches in the lawn. If you have a heavy infestation, a lot of severe damage can be done in a very short time. There are several species of sod webworm, and each part of the country has its own example. For instance the tropical sod webworm is a native of Florida and the burrowing sod webworm inhabits parts of Kansas, south to Louisiana and east to Maryland.

SOUTHERN BENT

If you dig into the drying sod and discover over four worms in a six inch area you'd better act. You can apply a soil insecticide or use milky spore disease dust which is an organic control (see organic controls for a further discussion of this control).

INSECTS THAT SUCK PLANT JUICES

CHINCH BUGS

The major damage is done by the nymphs or young bug. Small yellow spots appear on the lawn, soon becoming brown. The nymphs hatch from eggs laid by the female adults. You cannot see the first stage of the nymph without a microscope for it is the size of a pinhead. The first stage is a bright nymph with a white band across its back. After shedding its skin four times, the nymph is full grown. It is black and has a white spot on the back between the wing pads. The full grown chinch bug is black, about one sixth of an inch long, with white markings (see drawing for various stages of development).

Like other insects, this bug has various species for various regions of the country: Blissus hirtus, a hairy chinch bug that lives in the Eastern part of the country; B.leucopterus, a pest of bermudagrass of Oklahoma: B. insularis, a southern species that does a lot of damage to St. Augustinegrass in Florida and the Gulf States. The nymphs hatch in the spring and infest lawns until frost or cold weather.

REDTOP

Chinch bug

Chinch bug leaving mature wheat to feed on corn plant.
A. Life stages of the insect; a. eggs in wheat sheath; b. adult bug; c.red nymph; and d. black nymph.

The adults hibernate during the winter. In very warm climates like southern Florida they are active all winter except on cold days.

FALSE CHINCH BUG

Frequently mistaken for its namesake, which is a major pest but is not a problem. It may feed on grasses but rarely does heavy lawn damage. The nymphs are greenish gray and the adults are gray.

LEAFHOPPERS

Damage done by leafhoppers is often mistaken for disease or dry weather problems. The green grass turns a whitish color. These insects hop or fly from spot to spot and are tiny speckled, green or yellow bugs. There are many species that infest lawns and may cause serious damage to newly-seeded lawns--so much so that they may need to be reseeded.

MITES

The banks grass mite is found throughout the United States. Well-managed lawns that are healthy and well fertilized are rarely affected. As with all mites, these suck juices from the blades and the crowns.

The bermudagrass mite only does damage in the South. It is beige in color and very small. It causes short stems and rosetted growth. Clover mites feed on other plants in the lawn besides clover. They are tiny and red and seem to be more of a problem in the house than in the garden, as they can be carried into the house on the bottom of shoes or in loose cuttings on the clothes. They will infest shag rugs or furniture and prove a real nuisance. This usually occurs in the fall or spring.

SCALE INSECTS

The scale of grasses, like the scale of other plants, suck the juices of the plant. Some feed above the soil line on the blades of the grass, others feed below the surface on the roots. The grass becomes yellowish then turns brown. Severe damage is usually done during very hot, dry periods in the summer months. There are several kinds of scale that damage lawns in the country. Here are some of the more serious types:

RHODESGRASS SCALE

The adult is about one eighth of an inch in diameter and brownish-purple in color. They are covered with a cottony secretion. The small nymphs, or crawlers, move about in the first stages then settle down to feed in one spot. They secrete a wax that covers them. This scale may produce five generations in one season. The adults, the younger adults, and nymphs cause damage by sucking the crowns of the plant causing them to turn brown and die. The Rhodesgrass scale is found in Arizona, southern part of the West Coast, Hawaii, New Mexico, and the Gulf States. It attacks both St. Augustinegrass and bermudagrass.

BERMUDAGRASS SCALE

The adults are about one sixteenth of an inch long, oval and covered with a hard white secretion. It leaves brown patches and is very active in shady areas of the lawn.

GAMA GRASS

GROUND PEARLS (See photograph)

The female adult secretes a white wax sack, in which it puts about a hundred white-pink eggs. The tiny nymphs hatch and begin feeding on the tender parts of the grass. The nymphs cover themselves with a hard cover that looks like a pearl. Irregular brown spots in the lawn are the result of their feeding. They attack bermudagrass in the South and the Southwest, and Centipedegrass in the South.

SPITTLEBUGS

Not a serious threat to a well-kept lawn, but the nymphs, living in masses of spittle, do suck juices from the blades and crowns. The adults resemble the leafhopper but are a bit more active. Only two species are found in most home lawns: Meadow spittlebug--found in the Northeast and mid-West. The nymphs are yellowish-green. The adults are just over one quarter of an inch and are gray or brown. The adults feed on clover or weed plants; The two-lined spittlebug--found in the Southeast. This spittlebug in the nymph stage is white and its head is brown. The adult is about three-eighths of an inch long, is dark brown and has two orange stripes on its wings. They feed on centipedegrass, bermudagrass, and St. Augustinegrass.

INSECTS THAT INHABIT THE LAWN BUT DO NO DAMAGE

1. Chiggers: sometimes called redbugs, but are mites. If they get on your skin may cause severe irritation and itching.
2. Earwigs: breeds on grass clippings.
3. Fleas: my dog has fleas.
4. Millipedes and Centipedes: not harmful to lawns for they feed on decaying matter. They just look like they might eat you.
5. Slugs and Snails: a mucous trail on the walks indicates the presence of slugs or snails. You can trap them or poison them.
6. Sowbugs and Pillbugs: may feed on some grass but seldom a pest. They just roll away.
7. Thrips: usually just visiting from your flower bed. These tiny black or brown insects bite people and may be painful.
8. Ticks: usually fall from animals and can cause serious problems if carrying tick disease fever or tularemia.

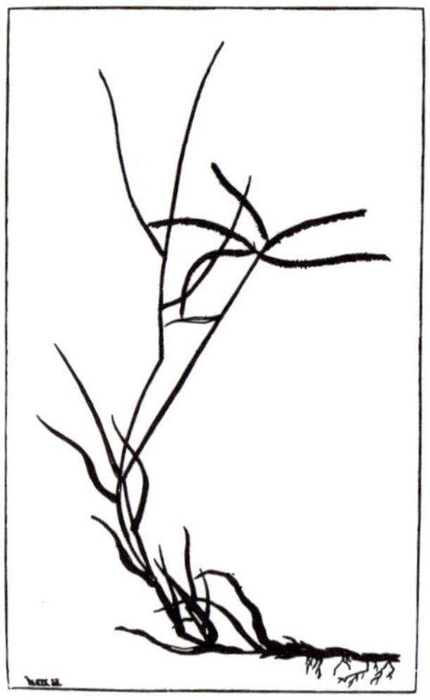

BERMUDA GRASSYARD GRASS

BROOM SEDGE

OTHER CAUSES OF POOR LAWNS

1. Grasses with an annual nature like annual rye will die after one season. It may reseed itself somewhat but you can expect brown instead of green in the spring.
2. Mixtures that do not blend well. Growing habits or different coloration.
3. Fertilizer burns when put on the lawn without a proper spreader or guide.
4. Oil spills or chemical spills.
5. Hydrated lime burn: use a fine grade agricultural lime in an even pattern.
6. Pesticide injury: Any chemical not applied in the proper manner will do damage to the lawn.
7. Dog urine: dogs seem to return to the same spot over and over, thus burning that spot or killing the grass.
8. Improper mowing: cutting too low in hot weather will burn the tops of the blades. Cut at about two inches to be safe.
9. Improper watering: frequent light watering induces shallow rooting of all grasses. Only water the lawn if it looks wilted. Never water more than a six inch soak of the soil.
10. Shallow top soil: you need a good ten inches for proper rooting.
11. Compacted soils: aerify your lawn. Hard rake it and punch holes in bad spots.

–Fertilizers for lawn use and application rate

Fertilizer	Type	Amount per 1,000 square feet when amount of nitrogen desired is—		
		1 pound	2 pounds	3 pounds
For establishment and maintenance				
		Pounds	*Pounds*	*Pounds*
5-10-5	Complete,[1] organic [2]	20	40	60
4-12-4	do	25	50	75
5-10-10	do	20	40	60
10-10-10	do	10	20	30
8-8-8	do	13	25	38
10-6-4	do	10	20	30
7-40-6	do	14	28	42
10-20-10	do	10	20	30
6-12-4	do	17	34	50
For maintenance				
		Pounds	*Pounds*	*Pounds*
Processed sewage sludge [3]	Nitrogen, organic [4]	17	34	51
Ammonium nitrate	Nitrogen, inorganic	3	6	9
Ammonium sulfate	do	5	10	15
Nitrate of soda	do	7	13	20
Steamed bonemeal	Nitrogen, phosphorus, organic.	50	100	150
Cottonseed meal	Nitrogen, organic	17	34	51
Peanut hull meal	do	50	100	150
Cocoa shell meal	do	50	100	150
Castor pomace	do	20	40	60
Dried cattle manure	Complete, organic	50	100	150
Dried sheep manure	do	70	140	210
Sewage sludge	Nitrogen, organic	50	100	150
Animal tankage	Nitrogen, phosphorus, organic.	15	30	45
Tobacco stems	Nitrogen, potash, organic	50	100	150
Urea	Nitrogen, inorganic	3	5	7
Processed tankage [5]	Nitrogen, organic	13	26	39
Soybean meal	do	17	34	51
Urea-form	Nitrogen, synthetic	3	6	9

[1] A complete fertilizer contains nitrogen, phosphorus, and potash.
[2] An inorganic fertilizer is a chemical fertilizer that is easily soluble and quickly available to plants.
[3] Milorganite and Huo-actinite are processed sewage-sludge materials.
[4] An organic fertilizer is slowly available to plants over a period of time.
[5] Agrinite is processed tankage material.

—Seeded grasses—rate and time of seeding

Grass	Pounds of seed per 1,000 square feet	Time of seeding
Bahiagrass	2-3	Spring.
Bermudagrass	2-3	Spring.
Blue gramagrass (unhulled)	1-1½	Spring.
Buffalograss (treated)	½-1	Spring.
Canada bluegrass	2-3	Fall.
Carpetgrass	3-4	Spring.
Centipedegrass	2-3	Spring.
Chewings fescue	3-5	Fall.
Colonial bentgrass (Highland, Astoria)	1-2	Fall.
Creeping bentgrass (Penncross, Seaside)	1-2	Fall.
Crested wheatgrass	1-2	Fall.
Japanese lawngrass (hulled)	1-2	Spring.
Kentucky bluegrass, common	2-3	Fall.
Kentucky bluegrass, Merion	1-2	Fall.
Red fescue	3-5	Fall.
Redtop	1-2	Fall.
Rough bluegrass	3-5	Fall.
Ryegrass (domestic and perennial)	4-6	Spring-fall.
Tall fescue (Alta, Ky. 31)	4-6	Fall.
Velvet bentgrass	1-2	Fall.
Mixture for sunny areas: 75% bluegrass, 25% red fescue	2-4	Fall.
Mixture for shady areas: 25% bluegrass, 75% red fescue	2-4	Fall.

SHEEP'S FESCUE.

GUIDE FOR SELECTING FUNGICIDES

Application Per 1,000 Square Feet—Follow Directions On Label

Disease and Casual Organism	Fungicide	Directions
Leafspot (Blight, Going-out, Melting-out) *Helminthosporium*	Acti-dione-thiram Captan Daconil 2787 Dyrene Fore Zineb	Disease can appear from April to August, depending on kind of grass and species of fungus. Treat your lawn every 7 to 14 days three times consecutively or until the disease has been controlled.
Brown Patch *Rhizoctonia solani*	Dyrene Fore PCNB	Disease can appear from June to August. Treat your lawn every 5 to 10 days until the disease has been controlled.
Rust *Puccinia*	Acti-dione-thiram Daconil 2787 Zineb	Disease can appear from June to September. Treat your lawn every 7 to 14 days until rust disappears.
Grease Spot and Cottony Blight *Pythium*	Dexon Zineb	Disease can appear from July to September and in fall and winter during warm, humid periods in the South. Treat your lawn every 5 to 14 days until the disease has been controlled.
Dollar Spot *Sclerotinia homeocarpa*	Acti-dione-thiram Daconil 2787 Dyrene Fore	Disease can appear from June to October. Treat your lawn at 7 to 14 day intervals until the disease has been controlled.
Stripe Smut Ustilago striiformis	Tersan 1991	Apply in October or early spring before grass begins growing. Water lawn well.
Snow Mold *Typhula* Fusarium Patch *Fusarium*	Dyrene Tersan 1991	Disease can appear from fall to spring. Treat your lawn at intervals of 2 to 6 weeks as needed.
Mushrooms Fairy Rings *Marasmius, Psalliota campestris, Lepiota*	Captan Methyl Bromide	Disease can appear throughout the growing season. Pour double or triple strength concentrate of captan into 1-inch holes punched 4 to 6 inches deep and 6 to 8 inches apart both inside and outside the affected area. Alternative method: fumigate infected area with methyl bromide; reseed or resod.
Slime Molds *Physarum cinereum*	Fore Zineb	Disease can appear throughout the growing season and can be controlled without fungicides. See page 9.

[1] Acti-dione-thiram formulations containing more than one active ingredient. Methyl bromide is applied as a vapor or gas. For all applications, read the manufacturer's directions.

CAUTION: Do not graze treated areas or feed clippings to livestock or poultry.

CONVERSION TABLE FOR SMALL QUANTITIES

1 level tablespoon = 3 level teaspoons
1 fluid ounce = 2 tablespoons
1 cup = 8 fluid ounces or 16 tablespoons
1 pint = 2 cups
1 quart = 2 pints or 32 fluid ounces
1 gallon = 4 quarts or 128 fluid ounces

Lawn pests and insecticides to use in controlling them

INSECTICIDE AND FORMULATION	DOSAGE OF FORMULATION PER 1,000 SQUARE FEET						
	GRUBS AND ANTS (*See note 1 below*)	SOD WEBWORMS, WIREWORMS, CICADA-KILLER WASP, AND WILD BEES (*See note 2 below*)	CHINCH BUGS AND FALSE CHINCH BUGS (*See note 3 below*)	ARMYWORMS, CUTWORMS, AND MOLE CRICKETS (*See note 4 below*)	EARWIGS, FIERY SKIPPER AND LUCERNE MOTH (*See note 5 below*)	CHIGGERS, FLEAS, AND TICKS	LEAFHOPPEF LEAF BUG, A MITES (*See note 6 belo*
GRANULES							
Carbaryl, 10 percent		2 pounds	2 pounds				
Chlordane, 5 percent	5 pounds	2½ pounds		2½ pounds	2 pounds		
Diazinon, 10 percent		2 pounds	2 pounds				
Ethion, 5 percent			5 pounds				
SPRAYS							
Wettable Powders							
Carbaryl, 50 percent		8 ounces	8 ounces	8 ounces	12 ounces	8 ounces	8 ounces.
Chlordane, 40 percent	10 ounces	5 ounces		5 ounces	4 ounces	2 ounces	2 ounces.
Diazinon, 50 percent		4 ounces	6 ounces	4 ounces		1¼ ounces	1¼ ounces.
Ethion, 25 percent			12 ounces				

Common and Scientific Names

Common name	*Scientific name*
Ants	Formicidae (many species).
Armyworms	Noctuidae (several species).
Asiatic garden beetle	*Maladera castanea.*
Banks grass mite	*Oligonychus pratensis.*
Bees	Hymenoptera (several species).
Bermudagrass mite	*Aceria neocynodomis.*
Bermudagrass scale	*Odonaspis ruthae.*
Billbugs	*Sphenophorus* spp.
Burrowing sod webworm	*Acrolophus* popeanellus.
Centipedes	Chilopoda.
Chiggers	*Trombicula* spp.
Chinch bug	*Blissus leucopterus.*
Cicada-killer wasp	*Sphecius speciosus.*
Clover mite	*Bryobia practiosa.*
Cutworms	Noctuidae (several species).
Earthworms	Annelida.
Earwigs	Dermaptera.
Eriophyid mite	*Aceria neocynodonis.*
European chafer	*Amphimallon majalis.*
European earwig	*Forficula auricularia.*
False chinch bug	*Nysius ericae.*
Fiery skipper	*Hylephila phyleus.*
Fleas	Siphonaptera (several species).
Fruit fly	*Oscinella frit.*
Grasshoppers	Orthoptera (many species).
Green June beetle	*Cotinis nitida.*
Ground pearls	*Margarodes meridionalis.*
Hairy chinch bug	*Blissus hirtus.*
Hunting billbug	*Sphenophorus venatus vestita.*
Japanese beetle	*Popillia japonica.*
Lawn armyworm	*Spodoptera mauritia acronyctoides.*
Leaf bug	*Spanogonicus albofasciatus.*
Leafhoppers	Cicadellidae (several species).
Lucerne moth	*Nomophila noctuella.*
May beetles	*Phyllophaga* spp.
Meadow spittlebug	*Philaenus spumarius.*
Millipedes	Diplopoda.
Mole crickets	Gryllidae (several species).
Northern masked chafer	*Cyclocephala borealis.*
Oriental beetle	*Anomala orientalis.*
Periodical cicada	*Magicicada septendecim.*
Rhodesgrass scale	*Antonina graminis.*
Rose chafer	*Macrodactylus subspinosus.*
Scorpions	Scorpionida.
Slugs and snails	Gastropoda.
Sod webworms	*Crambus* spp.
Southern chinch bug	*Blissus* insularis.
Southern masked chafer	*Cyclocephala immaculata.*
Sowbugs and pillbugs	Crustacea.
Spiders	Araneida.
Thrips	Thysanoptera (several species).
Ticks	Acarina.
Tropical millipede	*Orthomorpha corarctata.*
Tropical sod webworm	*Pachyzancla phacopteralis.*
Two-lined spittlebug	*Prosapia bicincta.*
Wireworms	Elateridae (several species).

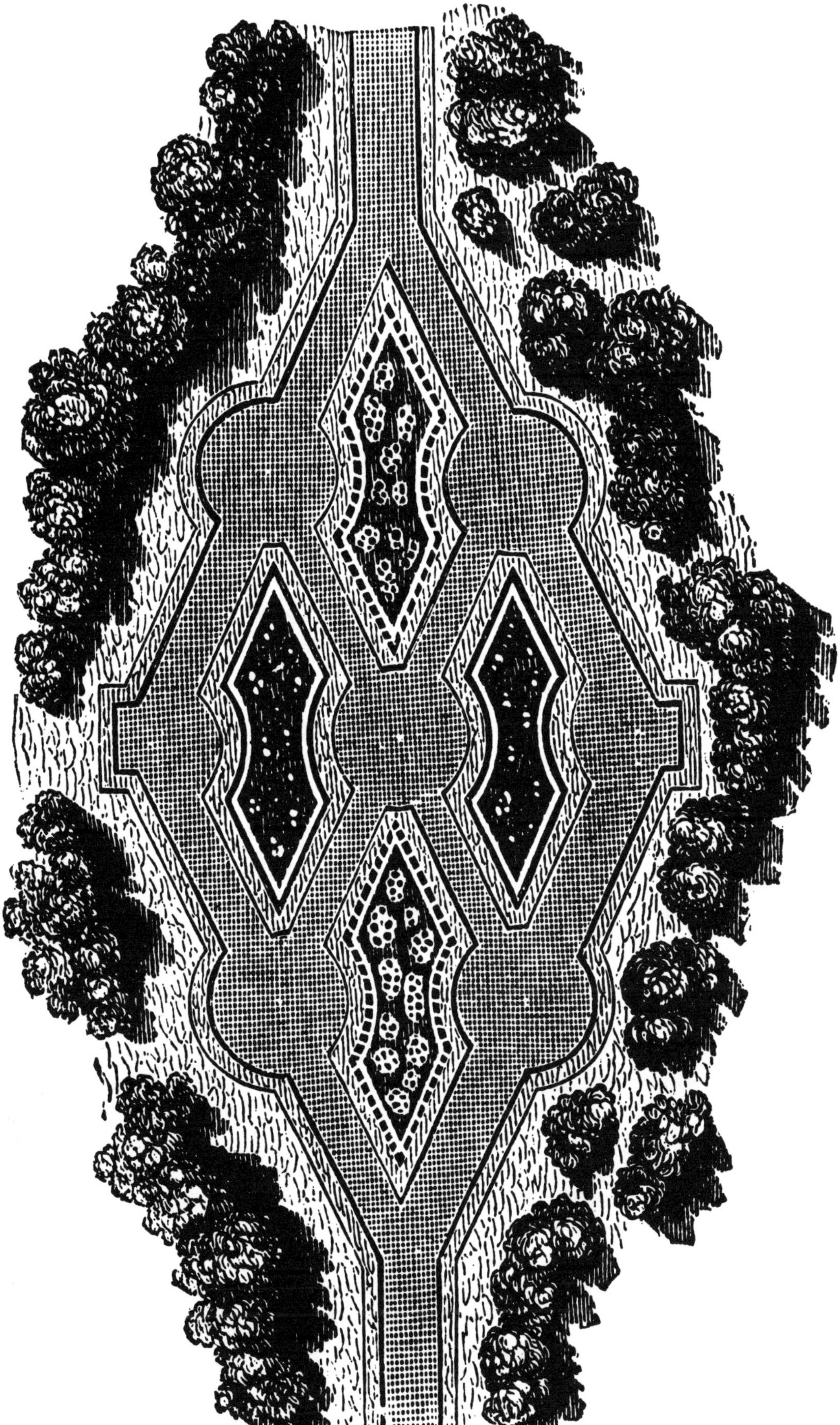

DESIGN FOR A FLOWER-BED IN A WALK.

AIR POLLUTION DAMAGE TO PLANTS

TREES

SHRUBS

AIR POLLUTION DAMAGE

Not only do chemicals in the air directly kill plants, they make them more susceptible to diseases and insect damage by causing a change in the growth patterns of certain plants, thus leaving them open for natural attacks that normally the plant would resist. Up to this time, not enough work has been done on the chronic injury of plants due to air pollution, and the long range results are not known. What I can safely say is that many symptoms that are passed off as diseases are in fact directly caused by pollution. It has been well documented that with the rise of technology and the increase in air pollutants, many of our native plants have been disappearing.

The major air pollutants are:

1. Ozone

2. Fluorides

3. Sulfur dioxide

4. Hydrogen chloride

5. Peroxyacyl nitrates (PAN'S)

6. Ethylene

7. Chlorine

8. Nitrogen oxides

THE SOURCE OF THE ABOVE POLLUTANTS

1 Ozone: automobile exhausts and combustion of fossil fuels.

2. Fluorides: a wide range of manufacturing such as brick making, glass factories, fertilizer plants, steel factories, and ceramic plants. Sensitive plants may be affected if air concentrations for a period of one day exceed one part per billion (PPB).

3. Sulfur dioxide: burning fossil fuels, paper mills, and power generating plants.

4. Hydrogen chloride: same as nitrogen oxide.

5. Peroxyacyl nitrates (PAN'S) same as ozone.

6. Ethylene: same as nitrogen oxide.

7. Chlorine: accidents in water purification stations, bleaching of flour, glass manufacturing, and burning of electrical wire to recover the copper.

8. Nitrogen oxide: fertilizer manufacturing, burning of soft coal, carbon arc operations, any kind of explosives.

THUYA GIGANTEA.

PLANTS SENSITIVE TO
PHOTOCHEMICAL SMOG

SHRUBS AND TREES

Acer (Maple)
Alnus (Alder)
Calycanthus (Carolina Allspice)
Ficus (Fig)
Gleditsia (Locust)
Hibiscus
Juglans (Walnut)
Mentha (Mint)
Petunia
Persea (Avacado)
Pinus (Pine)
Platanus (Sycamore)
Rhododendron
Robinia
Salix (Willow)
Salvia
Ulmus (Elm)

PLANTS RESISTANT TO
PHOTOCHEMICAL SMOG

SHRUBS AND TREES

Acacia
Aralia
Arbutus
Buxus (Boxwood)
Camellia
Cedrus (Cedar)
Cistus
Cotoneaster
Cupressus (Cypress)
Fraxinus (Ash)
Ginkgo (Maidenhairtree)
Prunus
Pittosporum
Pyracantha (Firethorn)
Quercus (Oak)
Spiraea (Bridal Wreath)
Syringa (Lilac)
Viburnum (in variety)
Yucca

PIN OAK. Quercus palustris

At this time it is impossible to diagnose long term chronic injury. The plants are suffering in a bio-chemical way and there may not be any outward signs to determine if a plant is in poor health. Leading experimenters in the field think premature dropping of foliage or premature color change may be caused by phytotoxicants.

When a plant is suffering acute damage due to air pollution, we now understand many symptoms that were not recognized just a few years ago. People living near or in the big cities like New York City have problems. As a former designer of terraces and penthouses in New York City, I have seen severe damage done by pollution. For a long time I admit being puzzled by clean holes in the foliage, and tip burns of certain evergreens, but as the information became more available the injury of these plants finally became evident.

OZONE: may cause chorotic flecking of the upper parts of the leaf and tip necrosis of evergreen needles.

SULFUR DIOXIDE: causes chlorosis of broadleaf evergreens such as rhododendron and mountain laural, and tip necrosis of various evergreen plants.

FLUORIDE: causes tip necrosis of evergreen needles, tip and margin necrosis of gladiolus.

If you want healthier plants and a cleaner air space you should join in the fight to clean up the mess that American industry, American cities and the general public have all contributed to.

—Lath screen for protecting plants

INDEX

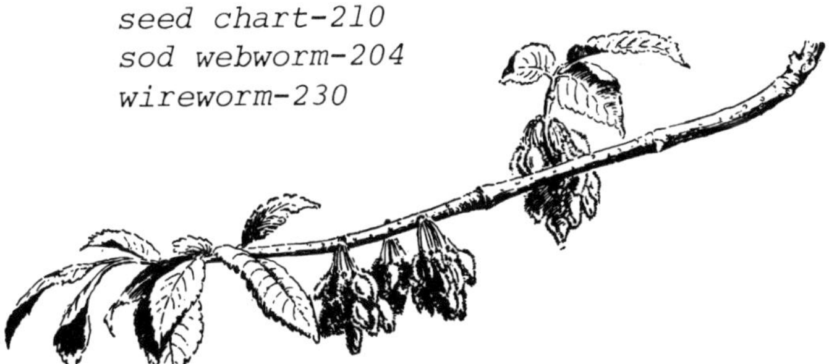

RUBUS (RASPBERRY, THE HANSELL).

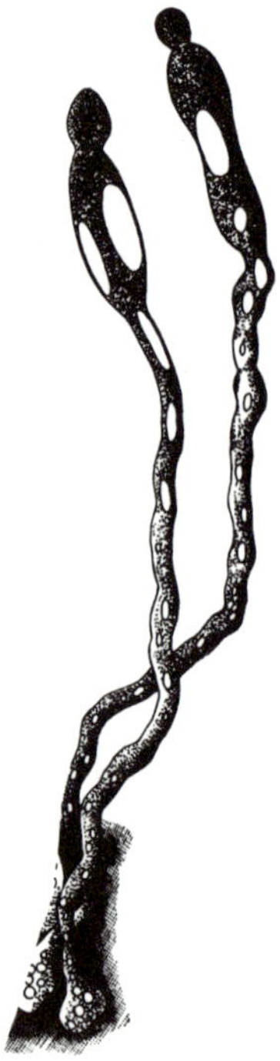

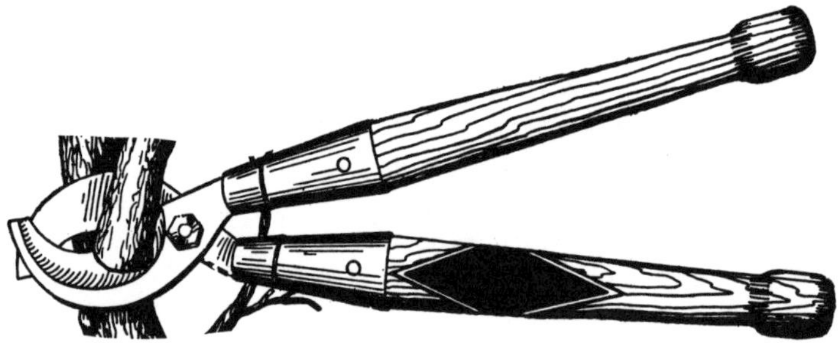

GARDEN NOTES

DISEASES PRESENT

INSECTS PRESENT

YEARLY CONTROL PROGRAM

FRUIT TREES

BRAMBLES

MINOR SHRUBS

GRAPES

GARDEN FLOWERS

LAWNS

PALMS

HOUSE PLANTS